Selected Titles in This Series

170 **Viktor Prasolov and Yuri Solovyev,** Elliptic functions and elliptic integrals, 1997
169 **S. K. Godunov,** Ordinary differential equations with constant coefficient, 1997
168 **Junjiro Noguchi,** Introduction to complex analysis, 1997
167 **Masaya Yamaguti, Masayoshi Hata, and Jun Kigami,** Mathematics of fractals, 1997
166 **Kenji Ueno,** An introduction to algebraic geometry, 1997
165 **V. V. Ishkhanov, B. B. Lur'e, and D. K. Faddeev,** The embedding problem in Galois theory, 1997
164 **E. I. Gordon,** Nonstandard methods in commutative harmonic analysis, 1997
163 **A. Ya. Dorogovtsev, D. S. Silvestrov, A. V. Skorokhod, and M. I. Yadrenko,** Probability theory: Collection of problems, 1997
162 **M. V. Boldin, G. I. Simonova, and Yu. N. Tyurin,** Sign-based methods in linear statistical models, 1997
161 **Michael Blank,** Discreteness and continuity in problems of chaotic dynamics, 1997
160 **V. G. Osmolovskiĭ,** Linear and nonlinear perturbations of the operator div, 1997
159 **S. Ya. Khavinson,** Best approximation by linear superpositions (approximate nomography), 1997
158 **Hideki Omori,** Infinite-dimensional Lie groups, 1997
157 **V. B. Kolmanovskiĭ and L. E. Shaĭkhet,** Control of systems with aftereffect, 1996
156 **V. N. Shevchenko,** Qualitative topics in integer linear programming, 1997
155 **Yu. Safarov and D. Vassiliev,** The asymptotic distribution of eigenvalues of partial differential operators, 1997
154 **V. V. Prasolov and A. B. Sossinsky,** Knots, links, braids and 3-manifolds. An introduction to the new invariants in low-dimensional topology, 1997
153 **S. Kh. Aranson, G. R. Belitsky, and E. V. Zhuzhoma,** Introduction to the qualitative theory of dynamical systems on surfaces, 1996
152 **R. S. Ismagilov,** Representations of infinite-dimensional groups, 1996
151 **S. Yu. Slavyanov,** Asymptotic solutions of the one-dimensional Schrödinger equation, 1996
150 **B. Ya. Levin,** Lectures on entire functions, 1996
149 **Takashi Sakai,** Riemannian geometry, 1996
148 **Vladimir I. Piterbarg,** Asymptotic methods in the theory of Gaussian processes and fields, 1996
147 **S. G. Gindikin and L. R. Volevich,** Mixed problem for partial differential equations with quasihomogeneous principal part, 1996
146 **L. Ya. Adrianova,** Introduction to linear systems of differential equations, 1995
145 **A. N. Andrianov and V. G. Zhuravlev,** Modular forms and Hecke operators, 1995
144 **O. V. Troshkin,** Nontraditional methods in mathematical hydrodynamics, 1995
143 **V. A. Malyshev and R. A. Minlos,** Linear infinite-particle operators, 1995
142 **N. V. Krylov,** Introduction to the theory of diffusion processes, 1995
141 **A. A. Davydov,** Qualitative theory of control systems, 1994
140 **Aizik I. Volpert, Vitaly A. Volpert, and Vladimir A. Volpert,** Traveling wave solutions of parabolic systems, 1994
139 **I. V. Skrypnik,** Methods for analysis of nonlinear elliptic boundary value problems, 1994
138 **Yu. P. Razmyslov,** Identities of algebras and their representations, 1994
137 **F. I. Karpelevich and A. Ya. Kreinin,** Heavy traffic limits for multiphase queues, 1994
136 **Masayoshi Miyanishi,** Algebraic geometry, 1994
135 **Masaru Takeuchi,** Modern spherical functions, 1994
134 **V. V. Prasolov,** Problems and theorems in linear algebra, 1994
133 **P. I. Naumkin and I. A. Shishmarev,** Nonlinear nonlocal equations in the theory of waves, 1994
132 **Hajime Urakawa,** Calculus of variations and harmonic maps, 1993

(*Continued in the back of this publication*)

Elliptic Functions and Elliptic Integrals

Translations of
MATHEMATICAL MONOGRAPHS

Volume 170

Elliptic Functions and Elliptic Integrals

Viktor Prasolov
Yuri Solovyev

AMS AMERICAN MATHEMATICAL SOCIETY
Providence, Rhode Island

В. В. Прасолов
Ю. П. Соловьёв

ЭЛЛИПТИЧЕСКИЕ ФУНКЦИИ И ЭЛЛИПТИЧЕСКИЕ ИНТЕГРАЛЫ

Translated from the original Russian manuscript by D. Leites

1991 *Mathematics Subject Classification.* Primary 11–01, 14H52, 33E05.

Abstract. This book is devoted to geometry and arithmetic of elliptic curves and to elliptic functions with applications to algebra and number theory. It includes modern interpretations of some famous classical algebraic theorems such as Abel's theorem on lemniscate and Hermite's solution of the fifth degree equation by means of theta functions. The book is self-contained and assumes as prerequisites only the standard one-year courses in algebra and analysis.

Library of Congress Cataloging-in-Publication Data

Prasolov, V. V. (Viktor Vasil′evich)
Elliptic functions and elliptic integrals / Viktor Prasolov, Yuri Solovyev.
p. cm. — (Translations of mathematical monographs, ISSN 0065-9282 ; v. 170)
"Translated from the original Russian manuscript by C. Leites"—T.p. verso.
Includes bibliographical references and index.
ISBN 0-8218-0587-8 | ISBN 0-8218-1346-1 (Softcover)
1. Elliptic functions. I. Solov′ev, I͡U. P. (I͡Uriĭ Pavlovich) II. Title. III. Title: Ėllipticheskie funkt͡sii i ėllipticheskie integraly IV. Series.
QA343.P73 1997
515′.983—dc21 97-24309
CIP

∞ The paper used in this book is acid-free and falls within the guidelines
established to ensure permanence and durability.
Visit the AMS homepage at URL: http://www.ams.org/

12 11 10 9 8 7 6 5 4 3 29 28 27 26 25 24

Contents

Preface ix

Chapter 1. Geometry of Cubic Curves 1
§1.1. Addition of points on a cubic 1
§1.2. Lines and curves on the projective plane 8
§1.3. The tangents and inflection points 11
§1.4. Normal forms of the nonsingular cubic 16
§1.5. Singular cubics 20
§1.6. No nonsingular cubic admits a rational parameterization 22

Chapter 2. Elliptic Functions 25
§2.1. The topological structure of nonsingular cubics in $\mathbb{C}P^2$ 26
§2.2. The elliptic functions 29
§2.3. The Weierstrass function 32
§2.4. A differential equation for the Weierstrass function $\wp(z)$ 35
§2.5. A parameterization of the cubic with the help of the Weierstrass function 36
§2.6. The elliptic integrals 39
§2.7. Addition theorems for the elliptic integrals $F(\varphi)$ and $E(\varphi)$ 44
§2.8. The elliptic Jacobi functions 46
§2.9. The Weierstrass theorem on functions possessing an algebraic addition theorem 49

Chapter 3. Arcs of Curves and Elliptic Integrals 53
§3.1. Arcs of the ellipse and the hyperbola 53
§3.2. Division of arcs of the ellipse 55
§3.3. Curves with elliptic arcs 60
§3.4. Curves whose arc lengths can be expressed in terms of arc lengths of the circle 64

Chapter 4. Abel's Theorem on Division of Lemniscate 67
§4.1. Construction of a regular 17-gon. An elementary approach 69
§4.2. Construction of regular polygons. Elements of Galois theory 71
§4.3. The equation for the division of the lemniscate 78
§4.4. Proof of Abel's theorem on the division of the lemniscate 86
§4.5. Several remarks on Serret's curves 91

Chapter 5. Arithmetic of Cubic Curves 103
§5.1. Diophantus' method of secants. Second degree diophantine equations 104

§5.2. Addition of points on a cubic curve 111
§5.3. Several examples 115
§5.4. Mordell's theorem 119
§5.5. The rank and the torsion group of an elliptic curve 124

Chapter 6. Algebraic Equations 131
§6.1. Solving cubic and quartic equations 131
§6.2. Symmetric polynomials 133
§6.3. The Lagrange resolvents 134
§6.4. Roots of unity 137
§6.5. The Abel theorem on the unsolvability in radicals of the general quintic equation 140
§6.6. The Tschirnhaus transformations. Quintic equations in Bring's form 145

Chapter 7. Theta Functions and Solutions of Quintic Equations 149
§7.1. Definition of theta functions 149
§7.2. Zeros of theta functions 150
§7.3. The relation $\Theta_3^4 = \Theta_2^4 + \Theta_0^4$ 151
§7.4. Representation of theta functions by infinite products 152
§7.5. The relation $\Theta_1'(0) = \pi\Theta_0(0)\Theta_2(0)\Theta_3(0)$ 154
§7.6. Dedekind's η-function and the functions f, f_1, f_2 155
§7.7. Transformations of theta functions induced by transformations of τ 156
§7.8. The general scheme of solution of quintic equations 158
§7.9. Transformations of order 5 159
§7.10. The change of parameter $\tau \mapsto \tau + 2$ 160
§7.11. The change of parameter $\tau \mapsto -\frac{1}{\tau}$ 161
§7.12. The change of parameter $\tau \mapsto \frac{\tau-1}{\tau+1}$ 163
§7.13. Functions invariant with respect to the changes of parameter $\tau \mapsto \tau + 2$, $\tau \mapsto -\frac{1}{\tau}$ and $\tau \mapsto \frac{\tau-1}{\tau+1}$ 164
§7.14. Deduction of the modular equation 165
§7.15. Solving quintic equations 166
§7.16. The main modular function $j(\tau)$ 169
§7.17. The fundamental domain of $j(\tau)$ 170
§7.18. How to solve the equation $j(\tau) = c$ 173
§7.19. The functions invariant under the changes of parameter $\tau \mapsto \tau + 1$ and $\tau \mapsto -\frac{1}{\tau}$ 175
§7.20. The functions invariant with respect to the changes of parameter $\tau \mapsto \tau + 2$ and $\tau \mapsto -\frac{1}{\tau}$ 176

Bibliography 179

Index 183

Preface

In June of 1796 the Literature Gazette, published at that time in Jena, offered to its readers the following note (in German):

New Discoveries.

Every novice in geometry knows that it is possible to construct geometrically, i.e., by ruler and compass, various regular polygons, namely, a triangle, a pentagon, a 15-gon and the polygons one can obtain from each of these by consecutive doubling the number of its sides. This was known already in the time of Euclid and, it seems, that the reigning belief, starting from that time, is that the domain of elementary geometry does not surpass these limits: at least I do not know any successful attempt to expand it in this direction. Hence, the discovery that, apart from these regular polygons, it is possible to geometrically construct a multitude of other polygons, for example, a 17-gon, seems to me worth noting. This discovery is essentially a mere corollary of a far-reaching theory not completely finished yet. The moment this theory is completed it will be offered to the public.

C. F. Gauss from Braunschweig,
student of mathematics in Göttingen.

The theory was completed five years later and published by Gauss in the 7th section of the famous *Disquisitiones Arithmeticae* (*Arithmetical Studies*), which appeared in 1801. Gauss proved that if the number n of sides of a regular polygon is of the form $n = 2^a p_1 \cdots p_k$, where the p_i are distinct Fermat primes, i.e., prime numbers of the form $2^{2^m}+1$, then the polygon can be constructed by ruler and compass. In algebraic language this statement means that for the numbers n indicated the equation $x^n - 1 = 0$ is solvable in quadratic radicals.

The proof of Gauss' theorem is based on a neat algebraic theory which served as the cornerstone for Galois theory created thirty years after *Arithmetical Studies* was published.

In the 7th section of *Arithmetical Studies*, apart from the theory of division of the circle, i.e., the algebraic theory of circular functions, there is a short remark, also by Gauss, to the effect that the method he developed is also applicable to certain higher transcendental functions; in particular, to functions related with integrals of the form $\int \frac{dx}{\sqrt{1-x^4}}$.

This remark became a starting point for the studies of Abel, who in 1827 proved that for the same values of n as mentioned by Gauss, it is possible to divide Bernoulli's lemniscate by ruler and compass into n equal parts. To do that, Abel had to considerably improve Gauss' method and, what is most important, create a new mathematical discipline — the *theory of elliptic functions.*

The theory of elliptic functions and its geometric twin — the *theory of elliptic curves* — occupies one of the central places in mathematics having unified several of its branches. In spite of its senior age, the theory of elliptic functions and elliptic curves remains an alive and rapidly developing domain of mathematics; it is an inexhaustible source of techniques, problems, and conjectures for the researchers.

In the past decade elliptic functions and curves became the subject of close attention by experts in such nonclassical fields as algebraic topology and quantum field theory; quite recently with the help of the elliptic curve theory Fermat's Last Theorem was finally proved.

The main topics of this book are the geometry of cubic curves, elliptic functions and their properties, elliptic integrals, addition theorems for elliptic functions and integrals, arcs of algebraic curves expressible via elliptic integrals, Abel's theorem on lemniscate, arithmetic properties of elliptic curves, Mordell's theorem, theta functions, and solutions of equations of the fifth degree.

In other words, the book is an introductory course on the theory of elliptic functions and elliptic curves and is aimed at those who encounter this topic for the first time. However, we hope that the book will be of interest to the experts as well.

The book is based on three lectures written by one of us (Yu. S.) in 1991–1992 as part of lectures for students organized by the Moscow Mathematical Society. The material of the book was collected for the optional course given by the second author (V. P.) in 1992–1993 at the Independent University of Moscow.

In writing this book we used a vast selection of literature, both classical treatises and various modern papers. We were greatly influenced by a remarkable paper by M. Rosen from *American Mathematical Monthly* [**C14**] that contains a modern proof of Abel's theorem. We have also borrowed a lot of useful facts from wonderful books of Husemoller [**B10**], Koblitz [**B12**] and Stepanov [**B22**].

The book does not assume from the reader any knowledge beyond the limits of beginning courses of mathematics majors in universities and is oriented to the widest range of readers: students of mathematics and physics, teachers, and even high school students. We hope that, having been acquainted with the subject of this book, the reader will be able to feel the charm of the fine art that we experienced while deciphering the works of old masters.

While the book was being written V. Prasolov benefited from a grant from the Russian Fund for Basic Research (95-01-00846).

CHAPTER 1

Geometry of Cubic Curves

§1.1. Addition of points on a cubic

A plane algebraic curve is the set of points $(x, y) \in \mathbb{R}^2$ satisfying the equation $f(x, y) = 0$, where $f(x, y)$ is a nonzero polynomial.

On certain plane curves, there exist natural laws for addition of points. For example, such laws exist on any straight line and on the unit circle $x^2 + y^2 = 1$. To be able to add points of the line, one should fix a point O on it and then the *sum of the points* X and Y can be defined as the point Z such that $\overrightarrow{OZ} = \overrightarrow{OX} + \overrightarrow{OY}$.

It is natural to define the sum of the points $(\cos\alpha, \sin\alpha)$ and $(\cos\beta, \sin\beta)$ on the unit circle to be the point $(\cos(\alpha+\beta),\ \sin(\alpha+\beta))$. This law of addition of points can be geometrically interpreted as follows. Let E be the point $(1, 0)$, A and B arbitrary points of the unit circle. Let us draw through E the straight line parallel to the straight line AB; the newly drawn line intersects the circle at the point C. Let us define the *sum of the points* A and B to be C.

In this form this definition works for any *conic* (a second order curve). Namely, fix a point E on a conic and consider the point at which the straight line drawn through E parallel to AB intersects the conic for the second time to be the *sum of the points* A and B. The commutativity of the operation obtained is obvious; the point E serves as the zero element. To find the element $-A$, one should draw through A the straight line parallel to the tangent at E. Only the associativity

$$(A + B) + C = A + (B + C)$$

is unclear. To prove it, denote the points $A+B$ and $B+C$ by P and Q, respectively. The associativity is equivalent to the following statement: *If A, B, C, E, P and Q are points on the conic such that $AB \parallel EP$ and $BC \parallel EQ$, then $AQ \parallel CP$.*

This statement is a particular case of *Pascal's theorem* on a hexagon inscribed in a conic.

EXAMPLES. a) For the parabola $y = x^2$ with the fixed point $E = (0, 0)$ the sum of the points (x_1, y_1) and (x_2, y_2) is the point $(x_1 + x_2, y_1 + y_2 + 2x_1x_2)$.

b) For the hyperbola $x^2 - y^2 = 1$ with the fixed point $E = (1, 0)$ the sum of the points (x_1, y_1) and (x_2, y_2) is the point $(x_1x_2 + y_1y_2, y_1x_2 + y_2x_1)$. Under the parameterization $x = \cosh t$ and $y = \sinh t$ this addition corresponds to the addition of the parameter t.

A *cubic* is a plane algebraic curve $\sum_{i,j} a_{ij}x^iy^j = 0$, where the greatest value of $i + j$ is equal to 3. On any nonsingular cubic, there also exists a quite natural law of addition of points. (We will discuss in detail what a *nonsingular cubic* is in §1.3.) The law of addition of distinct points of a cubic can be defined as follows.

On a cubic, fix an arbitrary point E (it will turn out to be the zero element). In order to add the points A and B, draw the straight line AB. It intersects the

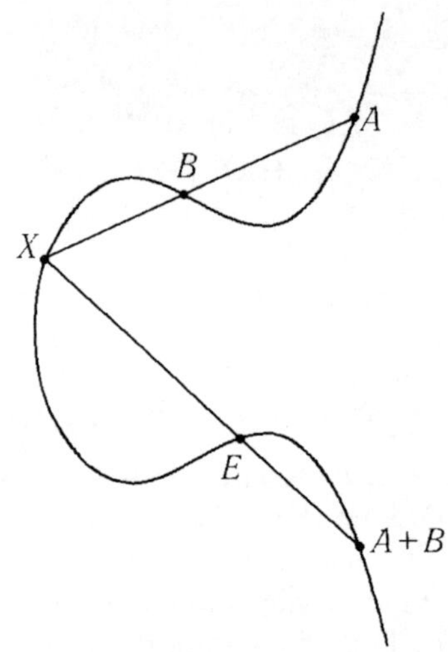

FIGURE 1

cubic at the point X. The point of intersection of the line XE with the cubic will be considered as the sum of A and B (Figure 1).

In the definition of the addition we used the following property of a cubic twice:

If a straight line intersects a cubic at two points, then the line intersects the cubic at precisely one more point.

This property seems to be almost obvious. Indeed, solve the equation of the straight line $ax + by + c = 0$ for x or y and substitute their values into the equation of the cubic. We get a third degree equation. By the hypothesis, it has two real roots and, therefore, there should exist a third real root.

In reality everything is not that simple. And the problem is not only that the polynomial can have multiple roots. The degree of the polynomial can also turn out to be lower than 3. In the latter case the operation of addition is degenerate: we cannot add any pair of points and this case is for now of no interest to us. We will discuss in §1.2 how it is possible to define addition for all points.

The commutativity of the obtained operation is obvious. It is also easy to verify that E is the zero element. The associativity of the operation is, however, not obvious. The equality $(A+B)+C = A+(B+C)$ is equivalent to the fact that the intersection points of the straight lines connecting the point $A+B$ with C and also $B+C$ with A lie on the cubic (Figure 2).

Denote the straight lines depicted in Figure 2 as follows:

$$p_1 = AB, \qquad p_2 = E(B+C), \qquad p_3 = C(A+B),$$
$$q_1 = BC, \qquad q_2 = E(A+B), \qquad q_3 = A(B+C).$$

Assume that all the intersection points of the straight lines p_i and q_j are pairwise distinct. Then the statement to be proved can be formulated as follows.

1.1.1. THEOREM. *Let A_{ij} be the intersection point of the straight lines p_i and q_j, where $1 \leq i, j \leq 3$ and the points A_{ij} are pairwise distinct. Suppose that it is known that all the points A_{ij}, except, perhaps, A_{33}, lie on a cubic. Then A_{33} also lies on this cubic.*

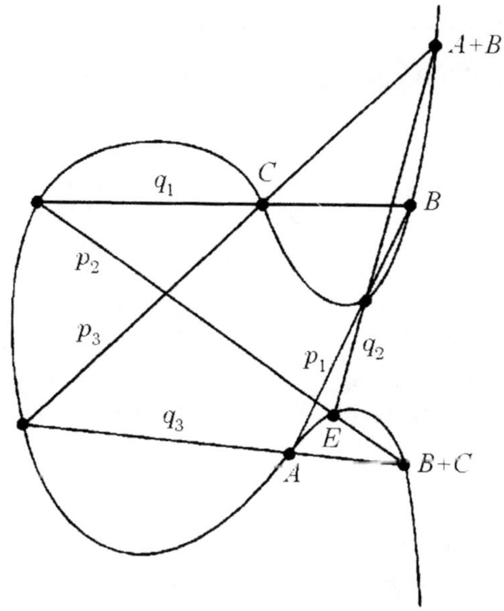

FIGURE 2

PROOF. Let $p_i(x, y) = 0$ and $q_j(x, y) = 0$ be the equations of the straight lines p_i and q_j. Then the third degree equation $p_1 p_2 p_3 = 0$ determines the triple of lines p_1, p_2 and p_3 and the equation $q_1 q_2 q_3 = 0$ determines the triple of lines q_1, q_2 and q_3. The cubic $\alpha p_1 p_2 p_3 + \beta q_1 q_2 q_3 = 0$ passes through all the points A_{ij}.

It turns out that one can represent in this way the equation of any cubic passing through eight of the nine points A_{ij}. Let us prove this.

Choose the straight lines p_1 and q_1 as coordinate axes, i.e., assume that $p_1(x, y) = y$ and $q_1(x, y) = x$. Let the given cubic be determined by the equation $P(x, y) = 0$. The functions $P(0, y)$ and $y p_2(0, y) p_3(0, y)$ vanish at the three points A_{11}, A_{21} and A_{31} on the y-axis (Figure 3). Moreover, these functions are polynomials of degree not higher than 3. Therefore, $P(0, y) = \alpha y p_2(0, y) p_3(0, y)$. Similarly,

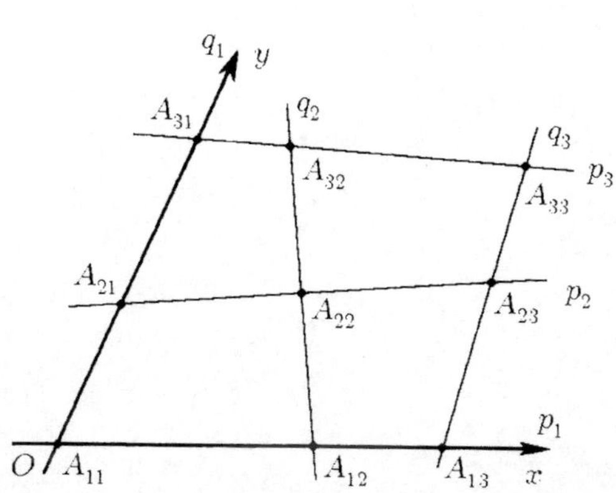

FIGURE 3

$P(x,0) = \beta x q_2(x,0) q_3(x,0)$. Consider the polynomial

$$Q(x,y) = P(x,y) - \alpha y p_2(x,y) p_3(x,y) - \beta x q_2(x,y) q_3(x,y).$$

Clearly,

$$Q(0,y) = P(0,y) = \alpha y p_2(0,y) p_3(0,y) = 0.$$

The polynomial $a_0(y) + a_1(y)x + a_2(y)x^2 + \cdots$ vanishes identically at $x = 0$ only if $a_0(y)$ is identically equal to zero, i.e., if this polynomial is divisible by x.

Similar arguments show that $Q(x,y)$ is divisible by y as well, i.e., $Q(x,y) = xyQ_1(x,y)$. The degree of $Q(x,y)$ does not exceed 3, hence, $Q_1(x,y)$ is either a linear function or a constant. Now, let us recall that the polynomials P, p_2p_3 and q_2q_3 vanish at the points A_{22}, A_{23} and A_{32} and, therefore, the polynomial Q also vanishes at these points. Since at all these points $xy \neq 0$, the linear function Q_1 must vanish at them. The points A_{22}, A_{23} and A_{32} do not lie on one line, and for a nonzero linear function f the equation $f(x,y) = 0$ determines a straight line. Hence, $Q_1 = 0$, i.e., $P = \alpha p_1 p_2 p_3 + \beta q_1 q_2 q_3$.

In particular, the point A_{33} lies on the curve $P(x,y) = 0$. We have also proved that any cubic passing through the points A_{ij} is given by the equation

$$\alpha p_1 p_2 p_3 + \beta q_1 q_2 q_3 = 0.$$

In other words, such curves constitute a one-parameter family. The proof of Theorem 1.1.1 is now completed and, together with it, the proof of associativity of the addition of points on a cubic. □

From Theorem 1.1.1, one can get a very simple proof of

PASCAL'S THEOREM. *The intersection points of opposite sides of an inscribed hexagon lie on one straight line* (Figure 4).

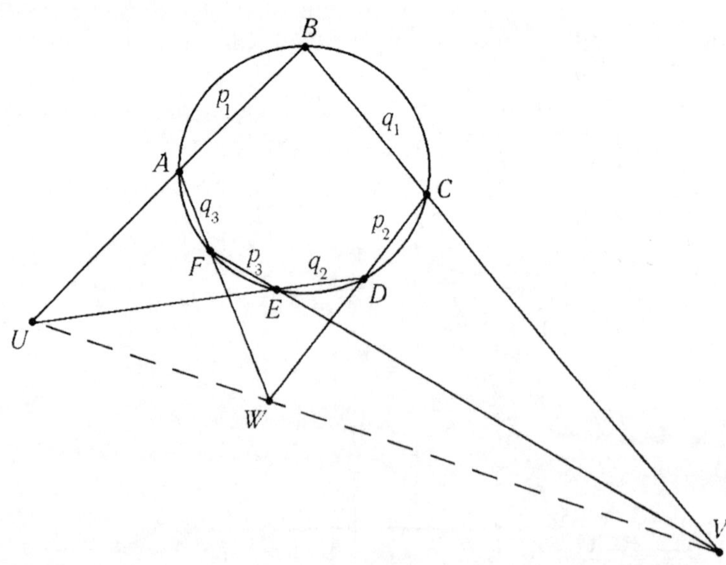

FIGURE 4

PROOF. Let $p_1 = AB$, $q_1 = BC$, $p_2 = EF$, $q_2 = DE$, $p_3 = CD$, $q_3 = AF$. As a cubic let us take the curve cut out by the equation $Ql = 0$, where $Q = 0$ is the equation of a circle and $l = 0$ is the equation of the straight line UV (here U and V are the intersection points of the lines p_1 with q_2 and p_2 with q_1, respectively). Let W be the intersection point of the lines p_3 and q_3. Of the remaining intersection points of the lines p_i and q_j it is known that they lie on the curve $Ql = 0$. The point W also lies on this curve and it must then lie on the straight line l, since it is not on the circle.

Instead of the circle $Q = 0$ one can take any second degree curve. In particular, we may assume that $Q = pq$, where p and q are linear functions. In this case we get

PAPPUS'S THEOREM. *Assume the points A, C and E on the straight line p as well as the points B, D and F on the straight line q are given. The lines AB and DE, BC and EF, AF and CD intersect at the points U, V, W, respectively (see Figure 5). Then the points U, V and W lie on one straight line.* □

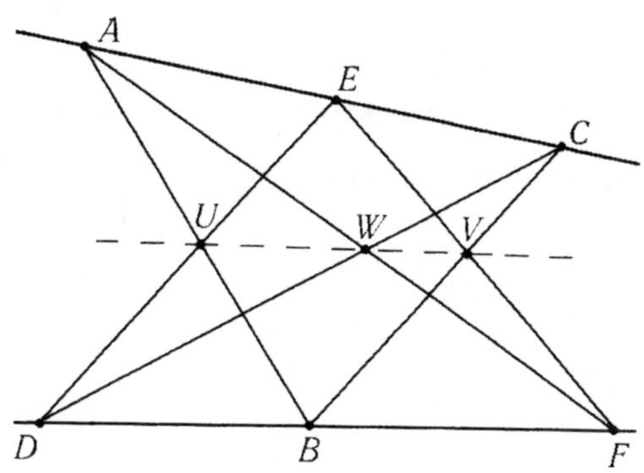

FIGURE 5

Sometimes we have to apply Theorem 1.1.1 when several of the points A_{ij} coincide. Hence, we have to understand how we should reformulate the theorem so that it remains true in such circumstances. In the course of the proof of the theorem we twice made use of the possibility to distinguish the points A_{ij}:

1) the function xy is nonzero at points A_{22}, A_{23} and A_{32} and, therefore, the linear function Q_1 vanishes at them;

2) these points do not lie on one straight line; hence, $Q_1 \equiv 0$. (Hereafter the sign $\equiv$ is sometimes used not as a congruence but to express the notion "is identically equal to".)

During the proof of Theorem 1.1.1 we only made use of the restriction of the polynomial P to the lines p_i and q_j. Hence, we may expect that instead of requiring that the points A_{ij} are distinct, it suffices to assume that

If two (or three) of the points A_{ij} on the line p_i or q_j coincide, then the restriction of the polynomial P to this line has at the point of coincidence a root of multiplicity two (or three).

This modification also concerns the point A_{33}.

Let us show that the formulation of Theorem 1.1.1 can indeed be modified in the way required. The proof of the fact that the polynomial $Q = P - \alpha p_1 p_2 p_3 - \beta q_1 q_2 q_3$

is divisible by $xy = p_1 q_1$ works without changes. If $A_{ij} = A_{ik} = A$, then at the point A the restriction of P to p_i has a root of multiplicity 2, the restriction of $p_1 p_2 p_3$ to p_i is identically zero and the restriction of $q_1 q_2 q_3$ has a root of multiplicity 2 because $q_j(A_{ij}) = 0$ and $q_k(A_{ik}) = 0$. Therefore, the restriction of Q to p_i has a root of multiplicity 2 at A.

The arguments are similar for the line q_j and also in the case of three coinciding points. Hence, it is clear that the linear function Q_1 still vanishes at points A_{22}, A_{23} and A_{32}.

If several of these points coincide, we use the fact that no nonzero linear function on the straight line can have a root of multiplicity 2.

As for the statement of Theorem 1.1.1, it is clear that for the restriction of the function $\alpha p_1 p_2 p_3 + \beta q_1 q_2 q_3$ to the line p_3 the multiplicity of the root at the point A_{33} is equal to the number of lines q_j passing through the point A_{33}.

For the line l tangent to the curve $F(X) = 0$ at a point X_0 the restriction of F to l is of multiplicity 2 at the point X_0. Indeed, let a point X_1 move along this curve towards the point X_0. The restriction of the function F to the straight line $X_0 X_1$ has roots X_0 and X_1. In the limit position the straight line $X_0 X_1$ coincides with l and the roots X_0 and X_1 merge into one root of multiplicity 2 (Figure 6(a)). The merging of three roots takes place on the tangent to the inflection point (Figure 6(b)). In §1.3 we will discuss in detail points of multiple intersection of a straight line with a cubic.

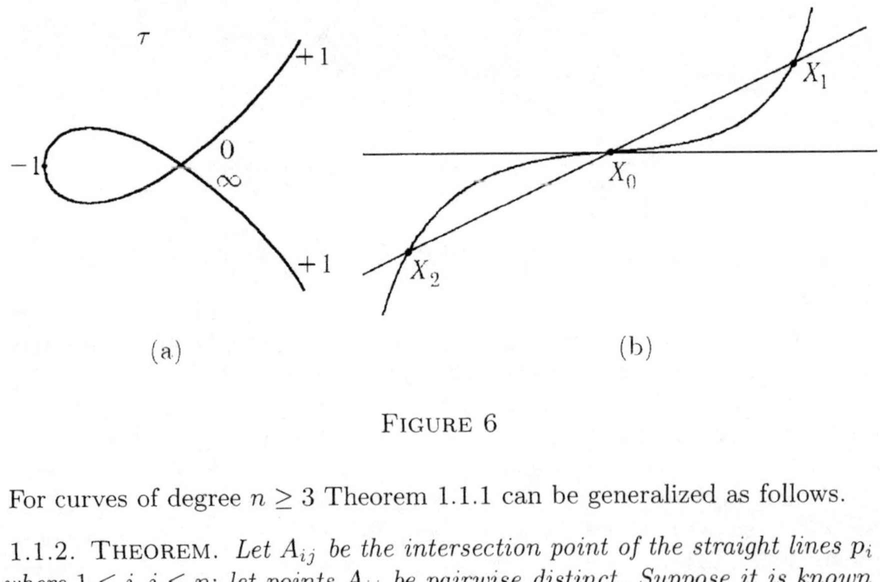

FIGURE 6

For curves of degree $n \geq 3$ Theorem 1.1.1 can be generalized as follows.

1.1.2. THEOREM. *Let A_{ij} be the intersection point of the straight lines p_i and q_j, where $1 \leq i, j \leq n$; let points A_{ij} be pairwise distinct. Suppose it is known that all points A_{ij}, where $i + j \leq n + 3$, lie on a curve of degree n. Then the other points A_{ij} also lie on this curve.*

PROOF. Let us take the straight lines p_1 and q_1 as coordinate axes. Let the curve in the formulation of the theorem be given by the equation $P_n(x, y) = 0$. Then $P_n(0, y) = \alpha p_1 \cdots p_n$ and $P_n(x, 0) = \beta q_1 \cdots q_n$. Consider the polynomial $Q_n = P_n - \alpha p_1 \cdots p_n - \beta q_1 \cdots q_n$. It suffices to demonstrate that $Q_n \equiv 0$. It is easy to verify that Q_n is divisible by $xy = p_1 q_1$, i.e., $Q_n = p_1 q_1 Q_{n-2}$. It remains to prove that the nonzero polynomial Q_{n-2} of degree not greater than $n - 2$ cannot

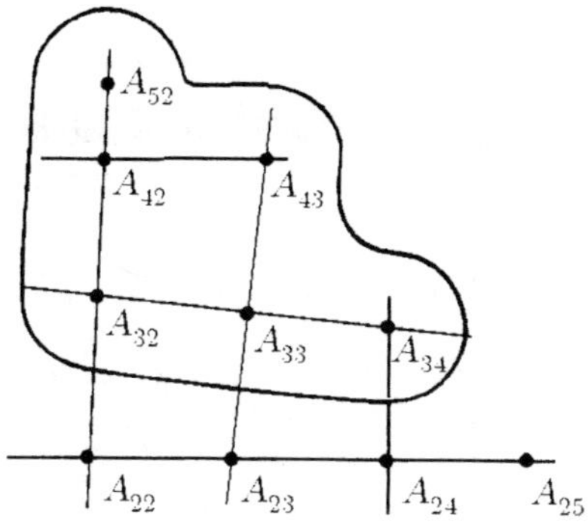

FIGURE 7

vanish at the points A_{ij}, where $i, j \geq 2$ and $i + j < n + 3$. (We shall prove this statement by induction on n.)

Suppose that such a nonzero polynomial Q_{n-2} exists. Its restriction to the straight line p_2 vanishes at $n-1$ points $A_{22}, A_{23}, \ldots, A_{2n}$. Therefore, the restriction of Q_{n-2} to this line is identically zero, i.e., $Q_{n-2} = p_2 Q_{n-3}$. The polynomial Q_{n-3} vanishes at points forming a similar configuration of lesser size (on Figure 7 these points are encircled with a thick curve).

These arguments illustrate the inductive step. The base of induction ($n = 3$) is considered in the proof of Theorem 1.1.1. □

* * *

A method of proving geometric theorems using the family of curves

$$p_1 p_2 p_3 + \mu q_1 q_2 q_3 = 0 \tag{1.1}$$

was developed by German mathematician **Julius Plücker** (1801–1868). The idea to represent a triple of lines as a degenerate cubic turned out to be quite fruitful. This representation allowed one to reduce the proof of various complicated geometric theorems to an ingenious selection of the coefficient μ in (1.1); this μ started to appear often in Plücker's papers.

Such an algebraization of geometry did not appeal to everybody. **Jacob Steiner** (1796–1863) — one of the most prominent geometers of that time — even flatly refused to ascribe signs to geometric quantities and preferred to consider instead distinct variants of the points' positions. In spite of the complicated way of treating the subject that Steiner chose, he succeeded several times in getting finer and deeper results than Plücker. Steiner referred unfavorably to new algebraic methods in geometry.

PROBLEMS

1.1.1. The straight lines AB and CD intersect at the point P, the straight lines BC and AD intersect at the point Q. A cubic passes through the points A, B, C, D, P, Q. Prove that the tangents to the cubic at the points P and Q intersect at a point that lies on the cubic.

HINT. Apply Theorem 1.1.1 in the case when $A_{31} = A_{32}$ and $A_{13} = A_{23}$.

1.1.2. A straight line intersects a cubic at the points A, B and C. The tangents to the cubic at the points A, B and C intersect the cubic at the points A_1, B_1 and C_1. Prove that the points A_1, B_1 and C_1 lie on one line.

HINT. Apply Theorem 1.1.1 in the case when $p_1 = p_2$.

1.1.3. An octagon with sides $l_1, \dots, l_8$ is inscribed in a conic. Prove that the eight intersection points of the lines l_i and l_j, where $j - i \equiv 3 \pmod 8$, lie on one conic.

HINT. Let $p_i = l_{2i-1}$, $q_i = l_{2i}$, $C_1 = 0$ be the initial conic, and $C_2 = 0$ be the conic passing through 5 of the 8 remaining intersection points of the straight lines p_i and q_j. Apply Theorem 1.1.2 to the curve $C_1C_2 = 0$.

1.1.4. Let the intersection points of the straight lines $p_1 = 0, \dots, p_n = 0$ be distinct. Prove that the equation of any curve of degree $n - 1$ passing through all intersection points is of the form

$$p_1 \cdots p_n \left(\frac{\lambda_1}{p_1} + \cdots + \frac{\lambda_n}{p_n} \right) = 0,$$

where $\lambda_1, \dots, \lambda_n$ are certain constants.

HINT. Let $C = 0$ be an equation of such a curve. Consider a straight line l not passing through the intersection points of the straight lines p_i. It is possible to select numbers λ_i so that at all n intersection points of l with the lines p_i we have

$$C - p_1 \cdots p_n \left(\frac{\lambda_1}{p_1} + \cdots + \frac{\lambda_n}{p_n} \right) = 0.$$

The same equality holds then at n points of any of the lines p_i.

§1.2. Lines and curves on the projective plane

In the preceding section we wrote that the addition of points on a cubic is defined, generally, not for all points. Let us illustrate this with an example of the curve

$$y^2 = x(x-1)(x-2) \tag{2.1}$$

plotted in Figure 8. Substituting the equation of the line $x = \frac{1}{2}$ into (2.1) we get $y^2 = \frac{3}{8}$. The degree of this equation is equal to 2, not 3. Hence, the line $x = \frac{1}{2}$ intersects the curve (2.1) at two points only, and the intersection points are not multiple ones. An attempt to add these points will not be successful.

If (x, y) is a point on the curve (2.1), then

$$\lim_{x\to\infty} \frac{x}{y} = \lim_{x\to\infty} \frac{x}{\sqrt{x^3}} = \lim_{x\to\infty} \frac{1}{\sqrt{x}} = 0.$$

A suspicion arises that both the curve (2.1) and the line $x = \frac{1}{2}$ pass through an infinite point in the direction of the y-axis. The lacking intersection point may turn out to be situated on the line at infinity.

Let us try to augment the collection of the points of the ordinary plane with points at infinity, thinking about these points as intersection points of parallel

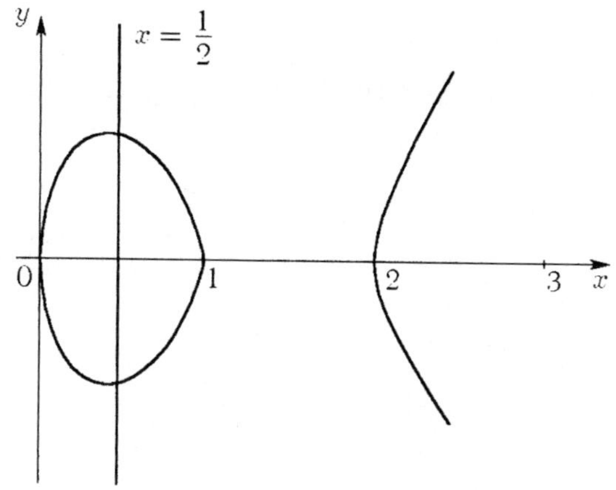

FIGURE 8

lines. Another reason for doing this is that otherwise even our formulations and proofs of Pappus's and Pascal's theorems would be inaccurate. Indeed, we have always assumed so far that the lines under consideration do indeed intersect. But they might be parallel as well. We can certainly consider, separately, the cases when certain lines intersect and certain lines are parallel, but this is quite tiresome because every case requires not only a separate formulation but a separate proof.

Put the plane π in a three-dimensional space and consider a point O outside the plane π. With every point $A \in \pi$ associate the straight line OA. To the line $l \in \pi$ this correspondence assigns not the whole plane Ol, i.e., the plane that contains the point O and the line l, but the part of Ol without the line l' that passes through O parallel to l. If the line $l_1 \in \pi$ is parallel to the line l, then the planes Ol and Ol_1 intersect along a straight line, l'. It is also clear that if the point A runs along the line l to infinity, then the limit position of the line OA is the straight line l'.

Define *the real projective plane* $\mathbb{R}P^2$ as follows. The *points* of $\mathbb{R}P^2$ are the lines passing through O. Let the *lines* in $\mathbb{R}P^2$ be the planes passing through O. In this picture the lines parallel to a plane π correspond to the infinite points of π and the plane parallel to π corresponds to the line at infinity in π. On the projective plane, any two straight lines intersect at one point. The projective lines corresponding to parallel lines in π intersect at a point of the line at infinity. When we forget about π, the infinite points do not differ from the other points.

To deal with algebraic curves we have to introduce coordinates on the projective plane. Assume that the point O is the origin of the coordinate system in a three-dimensional space and the plane π is given by the equation $z = 1$. A line passing through O consists of the points of the form $(\lambda x, \lambda y, \lambda z)$, where x, y, z are fixed and λ runs over $\mathbb{R}$. Therefore, we may consider the nonzero triples of real numbers (x, y, z) as *points* of $\mathbb{R}P^2$; here the triples (x, y, z) and $(\lambda x, \lambda y, \lambda z)$, $\lambda \neq 0$, are considered to be *equivalent*, and $\mathbb{R}P^2$ is the quotient of the set of triples modulo this equivalence relation. The line at infinity is given by the equation $z = 0$.

In the definition of the projective plane x, y, z and λ can be taken to be complex numbers. In this way we get a definition of the *complex projective plane*, $\mathbb{C}P^2$. The geometry of algebraic curves in $\mathbb{C}P^2$ is considerably simpler than that in $\mathbb{R}P^2$. This phenomenon is related to the fact that over $\mathbb{C}$ every nth degree polynomial has precisely n roots (multiplicities counted).

To the curve

$$y^2 = x(x-1)(x-2) \tag{2.2}$$

we can assign the curve

$$y^2 z = x(x-z)(x-2z) \tag{2.3}$$

on the projective plane. Indeed, equation (2.3) actually defines a curve on the projective plane because the points (x, y, z) and $(\lambda x, \lambda y, \lambda z)$ either simultaneously satisfy (2.3) or not. Moreover, on the plane π, given by the equation $z = 1$, both equations, (2.2) and (2.3), coincide.

Similarly, to any algebraic curve $\sum a_{ij}x^i y^j = 0$ we can assign the curve

$$\sum a_{ij}x^i y^j z^{n-i-j} = 0, \quad \text{where } n = \max(i+j),$$

on the projective plane. Now we can verify our hypothesis that the line $x = \frac{1}{2}z$ and the curve $y^2 z = x(x-z)(x-2z)$ meet at the infinite point in the direction of the y-axis. Substituting the expression $x = \frac{z}{2}$ into the equation of the curve we get $y^2 z = \frac{3z^3}{8}$. This equation has three types of solutions, namely, (1) $z = 0$, y is arbitrary; (2) $y = kz$ and (3) $y = -kz$, where $k = \sqrt{3/8}$ and z is arbitrary. In other words, each time we get a *family* of solutions, each family corresponding to one point of $\mathbb{C}P^2$.

Therefore, the line $x = \frac{1}{2}z$ on the projective plane does, indeed, intersect the curve considered at the three points: $(\frac{1}{2}, \sqrt{\frac{3}{8}}, 1)$, $(\frac{1}{2}, -\sqrt{\frac{3}{8}}, 1)$, $(0, 1, 0)$. The third point is the infinite one in the direction of the y-axis.

It is even possible to draw a sketch of what the line $x = \frac{1}{2}z$ and the curve considered look like in a neighborhood of the infinite point $(0, 1, 0)$. To this end, instead of the plane $z = 1$ one should take a plane passing through the point $(0, 1, 0)$ and not passing through the origin. Take, for example, the plane $y = 1$. On it, we get the curve $z = x(x-z)(x-2z)$. For small x and z our curve looks almost like the curve $z = x^3$ (Figure 9).

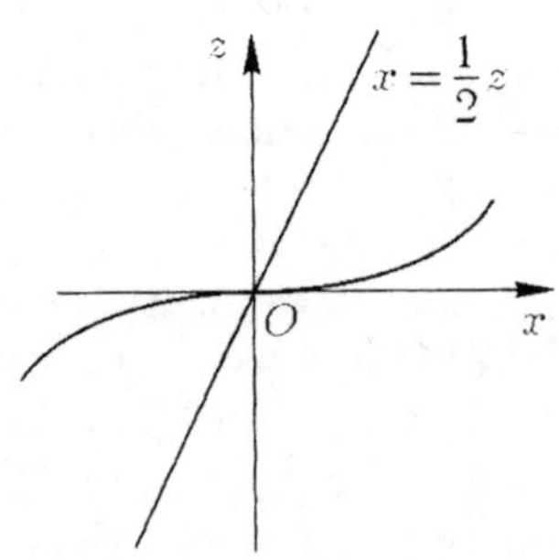

FIGURE 9

The passage to the projective plane is helpful not only in the case considered above. Let us show, for example, that any line on the projective plane either entirely belongs to the cubic or intersects it (multiplicities counted) at precisely three complex points; if we consider real points only, then it intersects the cubic at either one or three points.

We can find the intersection points of the straight line $ax + by + cz = 0$ and the cubic $\sum_{i+j+k=3} a_{ij}x^iy^jz^k = 0$ on the projective plane as follows. One of the numbers a, b, c is nonzero. Let, for instance, $c \neq 0$. Then $z = \alpha x + \beta y$, where $\alpha = -a/c$ and $\beta = -b/c$ (the case $\alpha = \beta = 0$ is not excluded). Inserting this expression into the equation of the cubic we get an equation of the form $Q = 0$, where $Q(x, y) = \sum b_p x^p y^{3-p}$. The following two cases are possible:

1) All the coefficients b_p are zero. Then the line $ax + by + cz = 0$ is entirely contained in the curve, i.e., Q is divisible by $ax + by + cz$.

2) Not all the coefficients b_p are zero. Then

$$Q(x, y) = bx^ry^s(x - t_1y)\cdots(x - t_my),$$

where $r + s + m = 3$. To the factor x^r there corresponds the intersection point $(0, 1, \beta)$ of multiplicity r; to the factor y^s there corresponds the point $(1, 0, \alpha)$ of multiplicity s; and to the factor $x - t_iy$ there corresponds the point $(t_i, 1, \alpha t_i + \beta)$.

A cubic polynomial Q with real coefficients can have either three real roots or one. Therefore, any line on the projective plane intersects a cubic either at three real points or at one point (multiplicities counted). Therefore, we have almost managed to clarify how to add noncoinciding points on a cubic.

We only have difficulties with points of self-intersection or with cusp points. The problem is that any line passing through such a point has a multiple intersection with the curve (cf. Problems 1.2.2 and 1.2.3). Therefore, taking the sum of such a point with any other point we never get any new points. We will discuss singular cubics in more detail in §1.5.

Now, it only remains to investigate the addition operation for coinciding points and figure out what the geometric meaning of the multiplicity of the intersection point is. We will do this in the following section.

Problems

1.2.1. Prove that the curve $y^2 = x^3 + px + q$ intersects the infinite line $z = 0$ at one point and the multiplicity of the intersection is equal to 3.

1.2.2. Prove that at the point $(0, 0)$ any line intersects the curve $y^2 = x^2(x+1)$ with multiplicity not less than 2 and for the lines $y = \pm x$ the multiplicity is equal to 3.

1.2.3. Prove that at the point $(0, 0)$ any line intersects the curve $y^2 = x^3$ with multiplicity not less than 2 and for the line $y = 0$ the multiplicity is equal to 3.

1.2.4. Prove that on the complex projective plane any circle $(x-a)^2+(y-b)^2 = R^2$ passes through the infinite points $(1, i, 0)$ and $(1, -i, 0)$.

§1.3. The tangents and inflection points

To add points A and B on a cubic we have to draw the line AB. How should one act if the points A and B coincide? Let us assume that the point A is fixed and the point B moves towards A along the given curve. Then, under certain conditions (point A should be nonsingular), the line AB tends to a fixed line, the tangent at A. Therefore, to find the sum $A + A$, instead of the line AB, we should draw the tangent at A (provided the tangent is uniquely defined at this point).

If the curve passing through points A and B is given by an equation $F = 0$, then the restriction of F to AB has roots at points A and B. In the limit position, when points A and B coincide, the restriction of F to A has a multiple root. Therefore, the restriction of F to the tangent has a multiple root at the tangent point. This property can be used in order to get the equation of the tangent.

Let the point $P = (p_1, p_2, p_3)$ belong to the curve $F = 0$, i.e., $F(P) = 0$, and let $X = (x_1, x_2, x_3)$ be an arbitrary point. The points of the projective line PX are of the form $\lambda P + \mu X$. The points of this line distinct from X are of the form $P + tX$. Let us consider the restriction of F to the line PX as a function of t. In the case of interest to us F is a polynomial of degree 3, hence,

$$F(P + tX) = F(P) + at + bt^2 + ct^3 = Q(t),$$

where $F(P) = 0$, $a = \sum F_i(P)x_i$, and $b = \frac{1}{2}\sum F_{ij}(P)x_ix_j$ (here F_i is the partial derivative of F with respect to the ith variable). The point P corresponds to the value $t = 0$. The polynomial $Q(t)$ has a multiple root at zero if $a = 0$, i.e., $\sum F_i(P)x_i = 0$.

A point P for which at least one of the numbers $F_i(P)$ is nonzero is called a *nonsingular point* of the curve. For a nonsingular point P the equation $\sum F_i(P)x_i = 0$ uniquely determines the *line l tangent to the curve* at P.

The tangent to the curve is geometrically uniquely defined regardless of the coordinate system. In our approach it is not yet clear that the definition of the tangent and the singularity of a point does not depend on the choice of the coordinate system. Let us prove the invariance of these definitions.

Now we show what happens under the change of coordinates $(x_1, x_2, x_3) \mapsto (u_1, u_2, u_3)$, where $x_i = \sum_j a_{ij}u_j$. Let $G(u_1, u_2, u_3) = F(x_1(u), x_2(u), x_3(u))$. Then

$$G_j = \frac{\partial G}{\partial u_j} = \sum_i \frac{\partial F}{\partial x_i}\frac{\partial x_i}{\partial u_j} = \sum_i F_i a_{ij}.$$

Since the matrix $J = (a_{ij})$ is nonsingular, the triple (G_1, G_2, G_3) is nonzero if and only if the triple (F_1, F_2, F_3) is nonzero. If f and g are rows (F_1, F_2, F_3) and (G_1, G_2, G_3), and x and u are columns $(x_1, x_2, x_3)^T$ and $(u_1, u_2, u_3)^T$, respectively, then $x = Ju$ and $g = fJ$. Hence $gu = (fJ)(J^{-1}x) = fx$ and the equations $fx = 0$ and $gu = 0$ determine the same line.

To pass from the projective coordinates (x_1, x_2, x_3) to the Cartesian ones (x_1, x_2) we have to set $x_3 = 1$. Assume $p_1 = 1$. To satisfy the condition $x_3 = 1$ for a point of the line P, we have to express the points of the line PX in the form $P + t(X - P)$. The expansion

$$F(P + t(X - P)) = \sum F_i(P)(x_i - p_i)t + \cdots$$

allows us to express the equation of the tangent in the form

$$F_1(P)x_1 + F_2(P)x_2 = F_1(P)p_1 + F_2(P)p_2.$$

In the projective coordinates, i.e., for a homogeneous function F, the expression $\sum F_i(P)p_i$ is equal to zero. The point is that for any homogeneous polynomial F of degree n the *Euler formula*

$$\sum F_i(X)x_i = nF(X)$$

holds. Consider a monomial $M = x_1^{p_1}x_2^{p_2}x_3^{p_3}$, where $\sum P_i = n$. Clearly, for the nonnegative p_i we have $x_i\partial_i(M) = p_iM$. It remains to recall that $p + q + r = n$.

Let P be a nonsingular point on the curve $F = 0$. Then the tangent l at the point P is defined. The restriction of F to l has a multiple root at P. If the multiplicity of this root is not less than 3, then P is called an *inflection point.* In other words, the condition $a = \sum F_i(P)x_i = 0$ must imply that $b = \frac{1}{2}\sum F_{ij}(P)x_ix_j = 0$, i.e., the quadric $\sum F_{ij}(P)x_ix_j = 0$ should contain the line $\sum F_i(P)x_i = 0$.

Recall that the second degree polynomial x^TAx (expressed here in the matrix form) is divisible by the linear function x^Tl only if $x^TAx = x^Tlm^Tx$ for some m. This means that the matrix $A = lm^T$ is the product of a column by a row, i.e., is of rank 1. In particular, $\det A = 0$. Thus, if P is an inflection point, then $\det(F_{ij}(P)) = 0$.

Let us show that for a nonsingular point on the curve the converse is also true, i.e., if P is a nonsingular point and $\det(F_{ij}(P)) = 0$, then P is an inflection point. Let us consider the quadric $\sum F_{ij}(P)x_ix_j = 0$. The point P belongs to it since by the Euler formula

$$\sum F_{ij}(P)p_ip_j = 2\sum F_j(P)p_j = 6F(P) = 0.$$

Moreover, the line $\sum_i F_i(P)x_i = 0$ is the tangent to this quadric at P. Indeed, the equation of the tangent to the quadric $\sum F_{ij}(P)x_ix_j = 0$ at P is of the form

$$\sum F_{ij}(P)x_ip_j = 0$$

and by the Euler formula $\sum_{i,j} F_{ij}(P)x_ip_j = 2\sum_i F_i(P)x_i$. We did not yet use the degeneracy of the quadric; in any case the tangent to the curve at P is at the same time the tangent to the quadric $\sum_{i,j} F_{ij}(P)x_ix_j = 0$. But in the case when this quadric consists of a pair of lines it entirely contains the tangent.

Let us summarize. The set of intersection points of the curves $F = 0$ and $H = 0$, where $H(X) = \det(F_{ij}(X))$, contains all the inflection points of the curve $F = 0$ (among these intersection points only singular points of this curve will not be inflection points). The curve $H = 0$ is called the *Hesse curve* or the *Hessian* of the curve $F = 0$. If F is a homogeneous polynomial of degree n, then F_{ij} is a homogeneous polynomial of degree $n - 2$. Therefore, H is a homogeneous polynomial of degree $3(n-2)$. For a cubic polynomial F the polynomial H is also a cubic one.

The invariance of the notion of the inflection point and of the Hesse curve can be proved almost in the same way as we proved the invariance of the tangent.

Let $G(u_1, u_2, u_3) = F(x_1(u), x_2(u), x_3(u))$, where $x_i = \sum a_{ij}u_j$. Then

$$G_{pq} = \frac{\partial^2 G}{\partial u_p \partial u_q} = \sum_{i.j} \frac{\partial^2 F}{\partial x_i \partial x_j}\frac{\partial x_i}{\partial u_p}\frac{\partial x_j}{\partial u_q} = \sum_{i.j} a_{ip}F_{ij}a_{jq},$$

i.e., $(G_{pq}) = J^T(F_{ij})J$, where $J = (a_{ij})$. Therefore, $\det(G_{pq}) = (\det J)^2\det(F_{ij})$. Hence, the conditions $\det(F_{ij}) = 0$ and $\det(G_{pq}) = 0$ are equivalent.

The search for the inflection points of the curve reduces to the search for the intersection points of the curve with the Hessian. So we have to find the intersection points of the two curves. We have already done this in the case when one of the curves is a straight line. The equation of the straight line enables us to express one variable in terms of the other one. Substituting this expression into the equation of the curve we can exclude one of the variables. For the curves of arbitrary degree we can also exclude one variable, but it is more difficult to do. To make the representation more visual, we will first consider curves in the Cartesian coordinates (x, y) and only afterwards pass to the projective coordinates (x, y, z).

For simplicity, let us confine ourselves to the third degree curves. It is possible to express the third degree polynomials $F(x, y)$ and $H(x, y)$ in the form

$$F(x, y) = a_0 y^3 + a_1(x) y^2 + a_2(x) y + a_3(x),$$
$$H(x, y) = b_0 y^3 + b_1(x) y^2 + b_2(x) y + b_3(x),$$

where $a_k(x)$ and $b_k(x)$ are polynomials of degree not greater than k, $0 \leq k \leq 3$. If (x_0, y_0) is a common point of the curves $F(x, y) = 0$ and $H(x, y) = 0$, then polynomials $f(y) = a_0 y^3 + a_1 y^2 + a_2 y + a_3$ and $h(y) = b_0 y^3 + b_1 y^2 + b_2 y + b_3$, where $a_k = a_k(x)$ and $b_k = b_k(x)$, have a common root y_0; the converse is also true: if the polynomials have a common root y_0, then the curves have a common point (x_0, y_0).

Over $\mathbb{C}$, two polynomials have a common root if and only if they have a common nonconstant divisor (over $\mathbb{R}$ the common divisor may have no roots). If $a_0 b_0 \neq 0$, then the polynomials $f(y)$ and $h(y)$ have a common divisor if and only if there exist polynomials h_1 and f_1 such that $f h_1 = h f_1$, where the degrees of h_1 and f_1 are lower than the degrees of $H(x, y)$ and $F(x, y)$, respectively.

Indeed, if f and h have a common divisor d, then we may set $f_1 = f d^{-1}$ and $h_1 = h d^{-1}$. If $f h_1 = h f_1$ and $\deg f_1 < \deg f$, then all prime divisors of f should occur in the prime factorization of $h f_1$; moreover, they enter with the same degrees. On the other hand, not all of them occur in the factorization of f_1.

The restriction $a_0 b_0 \neq 0$ is, no doubt, bothersome but in the projective case it is easy to satisfy.

Let $h_1(y) = u_0 y^2 + u_1 y + u_2$ and $f_1(y) = v_0 y^2 + v_1 y + v_2$. The equality $f h_1 = h f_1$ can be expressed in the form

$$\begin{array}{llllllll}
a_0 u_0 & & & -b_0 v_0 & & & = 0, \\
a_1 u_0 & +a_0 u_1 & & -b_1 v_0 & -b_0 v_1 & & = 0, \\
a_2 u_0 & +a_1 u_1 & +a_0 u_2 & -b_2 v_0 & -b_1 v_1 & -b_0 v_2 & = 0, \\
a_3 u_0 & +a_2 u_1 & +a_1 u_2 & -b_3 v_0 & -b_2 v_1 & -b_1 v_2 & = 0, \\
 & a_3 u_1 & +a_2 u_2 & & -b_3 v_1 & -b_2 v_2 & = 0, \\
 & & a_3 u_2 & & & -b_3 v_2 & = 0.
\end{array}$$

This system of linear homogeneous equations with respect to u and v has a nonzero solution if and only if its determinant vanishes, i.e.,

$$\begin{vmatrix}
a_0 & a_1 & a_2 & a_3 & & \\
 & a_0 & a_1 & a_2 & a_3 & \\
 & & a_0 & a_1 & a_2 & a_3 \\
b_0 & b_1 & b_2 & b_3 & & \\
 & b_0 & b_1 & b_2 & b_3 & \\
 & & b_0 & b_1 & b_2 & b_3
\end{vmatrix} = 0. \qquad (3.1)$$

This determinant is called the *resultant* of the polynomials f and h. The coefficients a_k and b_k depend on x and, therefore, the determinant (3.1) is a polynomial R of x, perhaps the zero one. For every root x_0 of the polynomial $R(x)$ the curves $F(x, y) = 0$ and $H(x, y) = 0$ have a common point (x_0, y_0). (Observe that in the real case the fact that $x_0 \in \mathbb{R}$ does not necessarily imply that $y_0 \in \mathbb{R}$.) If the polynomial $R(x)$ is the zero one, then the curves have a common component.

Now, let us repeat the above arguments for the projective coordinates. Expressing the polynomials F and H in the form

$$F(x,y,z) = a_0z^3 + a_1(x,y)z^2 + a_2(x,y) + a_3(x,y),$$
$$H(x,y,z) = b_0z^3 + b_1(x,y)z^2 + b_2(x,y) + b_3(x,y),$$

where a_k and b_k are homogeneous polynomials of degree k for $0 \le k \le 3$, it is possible to choose coordinates so that the curves $F = 0$ and $H = 0$ do not pass through the point $(0,0,1)$. Then the condition we need, $a_0b_0 \ne 0$, will be satisfied.

In the projective case the determinant (3.1) is a polynomial in two variables, $R(x,y)$. Let us prove that R is either the zero polynomial or a homogeneous polynomial of degree 9 (for curves of degrees m and n the degree of $R(x,y)$ is equal to mn). Indeed,

$$R(\lambda x, \lambda y) = \begin{vmatrix} a_0 & \lambda a_1 & \lambda^2 a_2 & \lambda^3 a_3 & & \\ & a_0 & \lambda a_1 & \lambda^2 a_2 & \lambda^3 a_3 & \\ & & a_0 & \lambda a_1 & \lambda^2 a_2 & \lambda^3 a_3 \\ b_0 & \lambda b_1 & \lambda^2 b_2 & \lambda^3 b_3 & & \\ & b_0 & \lambda b_1 & \lambda^2 b_2 & \lambda^3 b_3 & \\ & & b_0 & \lambda b_1 & \lambda^2 b_2 & \lambda^3 b_3 \end{vmatrix}.$$

Let us multiply the second and fifth rows by λ and the third and sixth by λ^2. As a result, we get a matrix for $R(x,y)$ in which the kth column is multiplied by λ^k. Therefore, $\lambda^6 R(\lambda x, \lambda y) = \lambda^{15} R(x,y)$, i.e., $R(\lambda x, \lambda y) = \lambda^9 R(x,y)$.

In the general case

$$\lambda^{p+q} R(\lambda x, \lambda y) = \lambda^r R(x,y),$$

where $p = 1+2+\dots+(n-1) = \frac{n(n-1)}{2}$, $r = 1+\dots+(m+n-1) = \frac{(m+n)(m+n-1)}{2}$ and $q = \frac{m(m-1)}{2}$.

It is easy to verify that $r - p - q = mn$.

The nonzero polynomial $R(x,y)$ can be represented in the form $\prod_{i=1}^{9}(y_ix - x_iy)$, where x_i and y_i do not vanish simultaneously. For every one of the nine pairs (x_i, y_i) there exists z_i such that (x_i, y_i, z_i) is the intersection point of the curves $f = 0$ and $h = 0$. The polynomial $R(x,y)$ can have multiple roots, i.e., certain pairs (x_i, y_i) can be proportional. Therefore, not every pair of cubics has nine distinct common points. But in the complex projective plane every two cubics have at least one common point. Hence,

any nonsingular cubic has an inflection point
(nine inflection points, multiplicities counted).

This is precisely the property we will need in the next section.

PROBLEMS

1.3.1. Prove that a point (x_0, y_0) on the curve $y = f(x)$ is an inflection point if and only if $f''(x_0) = 0$.

1.3.2. Prove that all points on the curve $y^2 = (x-x_1)(x-x_2)(x-x_3)$ are nonsingular if and only if the numbers x_i are distinct.

1.3.3. Prove that on the curve $y = (x - x_1)(x - x_2)(x - x_3)$ all points except $(0, 1, 0)$ are nonsingular.

1.3.4. Let A and B be inflection points on a cubic, and C the third intersection point of the straight line AB with the cubic. Prove that C is an inflection point.

HINT. Apply Theorem 1.1.1 to the case when $p_1 = p_2 = p_3 = AB$ and the lines q_1, q_2 and q_3 are the tangents at the points A, B and C, respectively.

1.3.5. The tangents to a cubic at the points A and B intersect at an inflection point P and the line AB intersects the curve at C. Prove that PC is a tangent to the curve.

HINT. Apply Theorem 1.1.1 to the case when p_1 is the tangent at P, $p_2 = p_3 = AB$, q_1 and q_2 are tangents at points A and B, and $q_3 = PC$.

§1.4. Normal forms of the nonsingular cubic

A cubic curve is called *nonsingular* if all its points are nonsingular. In this section we prove that over $\mathbb{C}$ the equation of a nonsingular cubic can be reduced by linear changes of homogeneous coordinates to any of the following forms:

1) $y^2z = x^3 + pxz^2 + qz^3$ (Weierstrass' form);
2) $x^3 + y^3 + z^3 = 3\lambda xyz$.

In the first case the polynomial $x^3 + px + q$ has no multiple roots (otherwise the curve is singular) and in the second case $\lambda^3 \neq 1$ (otherwise the curve consists of three lines).

Consider a nonsingular cubic $\sum a_{ij}x^iy^jz^{3-i-j} = 0$ over $\mathbb{C}$. In the preceding section we have shown that it has an inflection point. We may assume that the coordinates of the inflection point are equal to $(0, 1, 0)$ and the tangent to this point is given by the equation $z = 0$. In other words, the restriction of the function $F(x, y, z) = \sum a_{ij}x^iy^jz^{3-i-j}$ to the line $z = 0$ (i.e., the polynomial $a_{30}x^3+a_{21}x^2y+a_{12}xy^2 + a_{03}y^3$) has a root $x = 0$ of multiplicity 3. It follows that $a_{21} = a_{12} = a_{03} = 0$ but $a_{30} \neq 0$, since otherwise the curve considered would have contained the whole line $z = 0$. The tangent at $(0, 1, 0)$ is given by the equation

$$F_x(0,1,0)x + F_y(0,1,0)y + F_z(0,1,0)z = 0.$$

Hence, $F_x(0,1,0) = F_y(0,1,0) = 0$ but $F_z(0,1,0) \neq 0$, since otherwise the point $(0, 1, 0)$ would have been singular. The value of the homogeneous polynomial $F_z(x, y, z)$ of degree 2 at $(0, 1, 0)$ is equal to a_{02} and we may assume that $a_{02} = 1$. In Cartesian (not projective) coordinates the equation of the curve then takes the form

$$y^2 - 2(ax + b)y + P_3(x) = 0,$$

where P_3 is a third degree polynomial. Making the change of variables $y_1 = y - ax - b$ we get

$$y_1^2 - (ax + b)^2 + P_3(x) = 0,$$

i.e., $y_1^2 = Q_3(x)$, where $Q_3(x) = (ax + b)^2 - P_3(x)$ is a third degree polynomial. By a change of the form $x = \lambda x_1 + \mu$ the polynomial Q_3 can be reduced to the form $x_1^3 + px_1 + q$.

The polynomial Q_3 has no multiple roots, since otherwise the equation of the curve could have been reduced to the form $y^2 = x^2(\alpha x + \beta)$ and for such a curve the origin is a singular point.

In the preceding section we proved that any cubic has 9 inflection points, multiplicities counted, but we could not determine whether or not all of them are distinct. If the equation of the nonsingular cubic is expressed in the form $y^2 = Q_3(x)$, then it is easy to find the intersection points of the cubic with the Hessian and show that all of them are distinct.

1.4.1. THEOREM. *The nonsingular cubic $y^2 = Q_3(x)$ in $\mathbb{C}P^2$ has precisely 9 distinct inflection points.*

PROOF. We may assume that the polynomial Q_3 has a root $x = 0$, i.e., the curve considered is given by the equation $f(x, y) = 0$, where $f(x, y) = y^2 - x^3 - ax^2 - bx$. Since the polynomial Q_3 has no multiple roots, it follows that $b \neq 0$ and $a^2 - 4b \neq 0$. To get an equation of the Hessian, pass to homogeneous coordinates: $F(x, y, z) = y^2z - x^3 - ax^2z - bxz^2$. Then

$$\begin{aligned} H(x,y,z) &= \begin{vmatrix} -6x-2az & 0 & -2ax-2bz \\ 0 & 2z & 2y \\ -2ax-2bz & 2y & -2bx \end{vmatrix} \\ &= 8[(y^2+bxz)(3x+az) - (ax+bz)^2z], \end{aligned}$$

i.e., (dividing by 8) we have

$$h(x, y) = y^2(3x + a) + bx(3x + a) - (ax + b)^2.$$

It is easy to find the intersection points of the curves $f = 0$ and $h = 0$. Let us express the equation $f = 0$ in the form $y^2 = x^3 + ax^2 + bx$ and substitute this expression into the equation $h = 0$. As a result we get

$$(x^3 + ax^2 + bx)(3x + a) + bx(3x + a) - (ax + b)^2 = 0,$$

i.e.,

$$q(x) = 3x^4 + 4ax^3 + 6bx^2 - b^2 = 0.$$

Let us prove that the polynomial $q(x)$ has no multiple roots. Its derivative is equal to $12(x^3 + ax^2 + bx)$. Therefore,

$$q(x) - \frac{q'(x)}{12}\left(3x + a - \frac{b}{x}\right) = (4b - a^2)x^2.$$

Suppose that $q(x_0) = q'(x_0) = 0$. Then $x_0 \neq 0$, since $q(0) = -b^2 \neq 0$. On the other hand, $(4b - a^2)x_0^2 = 0$, where $4b - a^2 \neq 0$. Hence, $x_0 = 0$. Contradiction.

We have proved that the polynomial $q(x)$ has four distinct roots x_i. To every root x_i there correspond two distinct values of y because $y^2 = x_i^3 + ax_i^2 + bx_i = \frac{q'(x_i)}{12} \neq 0$. Thus, the curves $F = 0$ and $H = 0$ have 8 intersection points in the finite domain $z \neq 0$. Since $F(x, y, 0) = -x^3$ and $H(x, y, 0) = 24xy^2$, it follows that on the infinite line $z = 0$ the curves $F = 0$ and $H = 0$ have precisely one common point, $(0, 1, 0)$. The proof of Theorem 1.4.1 is completed. □

Any straight line passing through two inflection points contains one more inflection point. Indeed, we may assume that the coordinates of one of the inflection points are $(0, 1, 0)$. If (x_0, y_0) is a common point of the curve $y^2 = x^3 + ax^2 + bx$ and its Hessian $y^2(3x+a)+bx(3x+a) = (ax+b)^2$, then $(x_0, -y_0)$ is also a common point. The points $(0, 1, 0)$ and $(x_0, \pm y_0, 1)$ lie on the line $x = x_0z$.

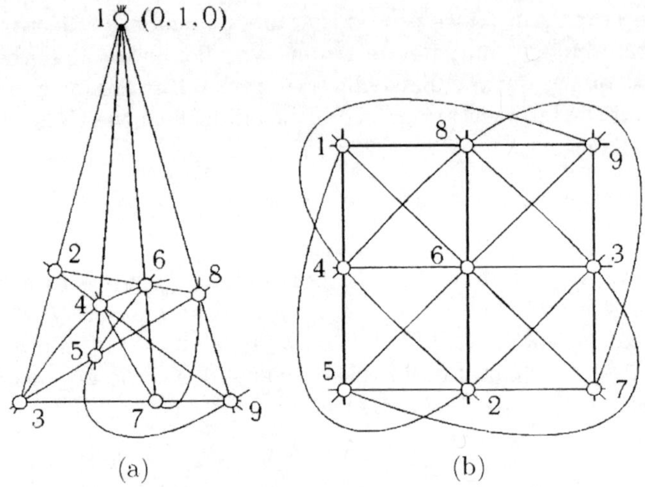

FIGURE 10

Schematically the configuration of the nine inflection points and the twelve straight lines that host them is depicted in Figure 10(a). A more symmetric scheme of this configuration is shown in Figure 10(b).

Recall one of the important notions of the projective geometry. Four points in $\mathbb{C}P^2$ are called *generic points* if no three of them lie on one straight line; four straight lines in $\mathbb{C}P^2$ are called *generic lines* if no three of them meet at one point.

Points with homogeneous coordinates $(x_i, y_i, z_i) = e_i$, where $i = 1, 2, 3, 4$, are generic if and only if the vectors e_1, e_2, e_3 are linearly independent and $e_4 = \lambda_1 e_1 + \lambda_2 e_2 + \lambda_3 e_3$, where $\lambda_1\lambda_2\lambda_3 \neq 0$. It is not difficult to show that *for any 4-tuple of generic points in $\mathbb{C}P^2$ there exists a projective transformation (i.e., a linear transformation of homogeneous coordinates) that maps this 4-tuple into any given 4-tuple of generic points.*

Indeed, let $\{e_i\}_{i=1}^4$ and $\{\varepsilon_i\}_{i=1}^4$ be two 4-tuples of generic points. Then $e_4 = \lambda_1 e_1 + \lambda_2 e_2 + \lambda_3 e_3$ and $\varepsilon_4 = \mu_1\varepsilon_1 + \mu_2\varepsilon_2 + \mu_3\varepsilon_3$, where $\lambda_1\lambda_2\lambda_3 \neq 0$ and $\mu_1\mu_2\mu_3 \neq 0$. The projective transformation required is as follows:

$$e_i \mapsto \alpha_i \varepsilon_i, \quad \text{where } \alpha_i = \frac{\mu_i}{\lambda_i} \text{ and } i = 1, 2, 3.$$

There is a one-to-one correspondence between the sets of points and the set of straight lines in $\mathbb{C}P^2$: to the point (a, b, c) there corresponds the line $ax+by+cz = 0$ (*projective duality*). If the points A and B lie on the line l, then the lines a and b dual to A and B meet at the point L dual to the line l.

To the line $ax+by+cz = 0$ there corresponds the point $(a, b, c) = \alpha$. Therefore, if $\mathbb{C}P^2$ is subjected to the projective transformation $(x, y, z) \mapsto (x, y, z)A$, where A is a nonsingular matrix, then the straight line $(x, y, z)\alpha^T = 0$ turns into the straight line $(x, y, z)A\alpha^T = 0$, i.e., $(x, y, z)(\alpha A^T)^T = 0$. Therefore, the transformation law of the coordinates on the line is $\alpha \mapsto \alpha A^T$. This transformation is also a projective one. Hence, the projective duality makes it possible to prove that any 4-tuple of generic lines can be turned into any other 4-tuple of generic lines by a projective transformation.

1.4.2. THEOREM. *By a change of coordinates the* 9 *inflection points of a cubic curve can be transformed into the following set of* 9 *points*:

$$\begin{matrix} (0,1,-1) & (0,\varepsilon^2,-\varepsilon) & (0,\varepsilon,-\varepsilon^2) \\ (-1,0,1) & (-\varepsilon^2,0,1) & (-\varepsilon,0,1) \\ (1,-1,0) & (-\varepsilon,1,0) & (-\varepsilon^2,1,0), \end{matrix}$$

where $\varepsilon^3 = 1$ *and* $\varepsilon \neq 1$, *i.e.*, $\varepsilon^2 + \varepsilon + 1 = 0$.

PROOF. The straight lines 189, 463, 527 (Figure 10(b)) cannot intersect at one point. Indeed, let R be a common point of these lines. Any 4-tuple of generic points in $\mathbb{C}P^2$ can be transformed by a projective transformation into any other four generic points. Therefore, we may assume that the points R, 1, 5 and 6 are real ones. Then all the other points of the configuration are also real. It is easy to see that this is impossible.

Adding the line 145 to the indicated triple of lines we get a 4-tuple of generic lines. Therefore, we may assume that the lines 145, 189, 463 and 527 are given by the equations $x + y + z = 0$, $x = 0$, $y = 0$ and $z = 0$, respectively. Then the coordinates of the inflection points are of the following form:

$$\begin{matrix} (0,1,-1) & (0,a,-b) & (0,c,-d) \\ (-1,0,1) & (-a',0,1) & (-c',0,1) \\ (1,-1,0) & (-b',1,0) & (-d',1,0), \end{matrix} \tag{4.1}$$

where all the numbers a, b, ... , d' are nonzero.

The points $(0, a, -b)$, $(-a', 0, 1)$, and $(-b', 1, 0)$ lie on one straight line, namely, on line 862. Therefore, $ab' = ba'$ and we may assume that $a = a'$ and $b = b'$. Similarly, $c = c'$ and $d = d'$.

Considering the lines 167 and 123 we get $a = d$ and $b = c$, respectively. Considering the lines 538 and 596 we get $a = bc$ and $c = ad$, respectively. It follows that $b^3 = 1$ and $a = b^2$. It is also clear that $b \neq 1$. Substituting $a = \varepsilon^2$, $b = \varepsilon$, $c = \varepsilon$ and $d = \varepsilon^2$ into (4.1) we get the points required.

It is easy to verify that the remaining eight lines containing triples of given points are given by equations of the form $x + \alpha y + \beta z = 0$, where α and β take values 1, ε and ε^2. □

With the help of Theorem 1.4.2 it is easy to prove that *any nonsingular cubic can be reduced to the form* $x^3 + y^3 + z^3 + 3\lambda xyz + 0$. Indeed, let the inflection point of the given curve have the coordinates indicated in the formulation of Theorem 1.4.2. These points belong to both the triple of lines $xyz = 0$ and the triple of lines

$$(x + y + z)(x + \varepsilon y + \varepsilon^2 z)(x + \varepsilon^2 y + \varepsilon z) = 0.$$

Therefore, any cubic passing through these nine points is given by the equation

$$\mu xyz + \nu(x + y + z)(x + \varepsilon y + \varepsilon^2 z)(x + \varepsilon^2 y + \varepsilon z) = 0.$$

It remains to note that

$$(x + y + z)(x + \varepsilon y + \varepsilon^2 z)(x + \varepsilon^2 y + \varepsilon z) = x^3 + y^3 + z^3 - 3xyz.$$

Problems

1.4.1. Prove that there is a linear change $x' = ax + b$ that reduces the curve

$$y^2 = (x - x_1)(x - x_2)(x - x_3)$$

to the form

$$y^2 = x'(x' - 1)(x' - \lambda), \quad \text{where} \quad \lambda = (x_3 - x_1) : (x_2 - x_1).$$

In this way we can get 6 distinct values of λ for the same curve, namely,

$$\lambda, \quad \lambda^{-1}, \quad 1 - \lambda, \quad (1 - \lambda)^{-1}, \quad (\lambda - 1)\lambda^{-1}, \quad \lambda(\lambda - 1)^{-1}.$$

1.4.2. Let $F = 0$ be the equation of a cubic, and $H = 0$ the equation of its Hessian. Prove that $\lambda F + \mu H = 0$ is the equation of a cubic with the same inflection points as the initial curve (assuming that the initial curve is nonsingular).

1.4.3. Prove that the curve $x^3 + y^3 + z^3 = 3\lambda xyz$ is nonsingular if and only if $\lambda^3 \neq 1$.

1.4.4. Prove that the curve $x^3 + y^2 z + axz^2 = 0$ is the Hessian of its own Hessian.

1.4.5. Prove that the curve $x^3 + y^3 + z^3 = 3\mu xyz$ is the Hessian of the curve $x^3 + y^3 + z^3 = 3\lambda xyz$ if and only if $\mu = -\frac{4+\lambda^3}{3\lambda^2}$.

§1.5. Singular cubics

The equation of a nonsingular cubic curve can be written in the form

$$y^2 = (x - x_1)(x - x_2)(x - x_3),$$

where the numbers x_1, x_2 and x_3 are distinct. In the real case such a curve is shown in Figure 11.

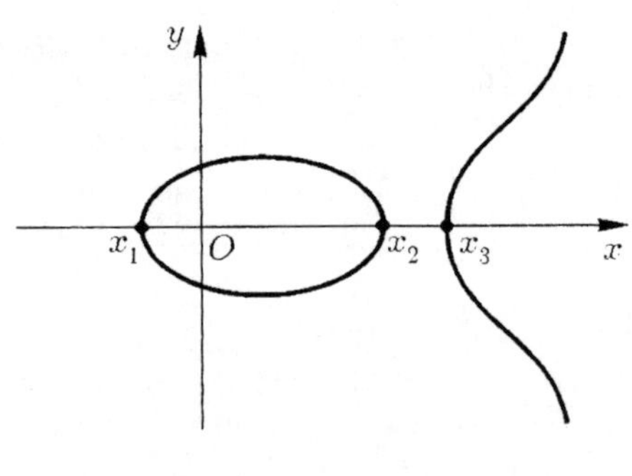

FIGURE 11

Let $x_1 < x_2 < x_3$. When the roots x_1 and x_2 merge, we get a curve of the form $y^2 = x^2(x - 1)$ (see Figure 12(a)); when the roots x_2 and x_3 merge, we get a curve of the form $y^2 = x^2(x + 1)$ (see Figure 12(b)). Over $\mathbb{R}$ these curves are distinct, but over $\mathbb{C}$ the distinction between them disappears.

If all the three roots merge we get the curve $y^2 = x^3$ (see Figure 12(c)). For all three curves the origin is a singular point.

Any straight line $y = kx$ intersects the curves $y^2 = x^2(x \pm 1)$ and $y^2 = x^3$ at the singular point with multiplicity at least two. Indeed, for the equations

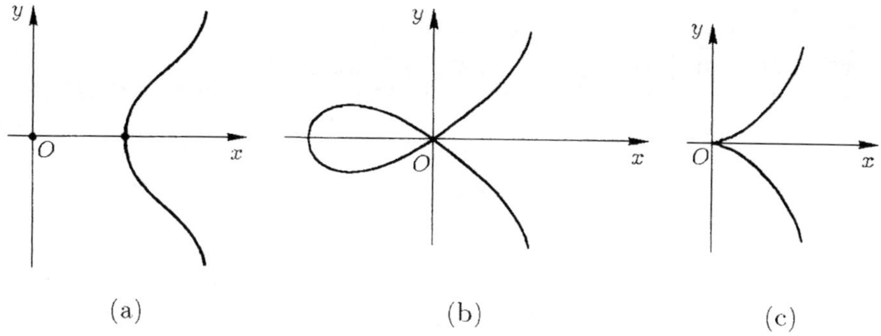

FIGURE 12

$k^2x^2 = x^2(x \pm 1)$ and $k^2x^2 = x^3$ the root $x = 0$ is at least a double one. Therefore, for any straight line connecting the singular point with another point on a cubic the third intersection point is the singular point. Therefore, the addition of the singular point to any other point should always give the singular point. Hence, we cannot define addition for the singular point. But if we exclude the singular point, then for either of the curves $y^2 = x^2(x+1)$ or $y^2 = x^3$ the addition of points is well defined. If for the zero element we take the infinite point in both cases, then the curve $y^2 = x^2(x+1)$ over $\mathbb{R}$ turns into the group of nonzero real numbers with respect to multiplication and the curve $y^2 = x^3$ turns into the group of real numbers with respect to addition.

(Over $\mathbb{C}$ we get the group $\mathbb{C} \setminus \{0\}$ with respect to multiplication and $\mathbb{C}$ with respect to addition, respectively.)

Let us start with the curve $y^2 = x^3$. This curve admits a rational parameterization $x = t^{-2}$, $y = t^{-3}$. The intersection points of this curve with the line $ax + by + c = 0$ are determined by the relation $ct^3 + at + b = 0$. If the line does not pass through the singular point, then $c \neq 0$. In this case we get a cubic equation with the zero coefficient of t^2. The sum of the roots of such an equation is equal to zero: $t_1 + t_2 + t_3 = 0$. Let us take for the zero element E the infinite point corresponding to the parameter $t_E = 0$. Assume t_A and t_B to be the values of parameter corresponding to the points A and B of the given curve. The straight line AB intersects the cubic at a point X; we have $t_A + t_B + t_X = 0$. The line EX intersects the curve at the point $A + B$, i.e., $t_E + t_X + t_{A+B} = 0$. Hence, $t_{A+B} = -t_X = t_A + t_B$. Hence to add points on the curve $y^2 = x^3$, we must add the corresponding values of the parameter t. Observe that to the singular point there corresponds the value $t = \infty$ of the parameter.

The curve $y^2 = x^2(x+1)$ also admits a rational parameterization. Indeed, let $y = tx$. Then $t^2x^2 = x^2(x+1)$, i.e., $x = t^2 - 1$ and $y = tx = t^3 - t$. The straight line $ax + by + c = 0$ intersects the curve $y^2 = x^2(x+1)$ at the point whose value of parameter satisfies the relation

$$a(t^2 - 1) + b(t^3 - t) + c = 0.$$

If $b \neq 0$, then after the division by b we get a cubic equation with coefficient 1 of t^3 and -1 of t. The roots of such an equation satisfy the relation $t_1t_2 + t_2t_3 + t_3t_1 = -1$.

A simpler relation can be obtained after the reparameterization

$$t = (1+\tau)(1-\tau)^{-1}.$$

Indeed, it is easy to verify that $\tau_1\tau_2\tau_3 = 1$. For the zero element E take the infinite point corresponding to the value of parameter $\tau_E = 1$. To find the sum $A + B$, we must consider the point X at which the straight line AB intersects the cubic. Since $\tau_A\tau_B\tau_X = 1$ and $\tau_X\tau_E\tau_{A+B} = 1$, it follows that $\tau_{A+B} = \tau_A\tau_B$. When we add points of the curve $y^2 = x^2(x+1)$, we multiply the corresponding values of parameter τ. To the singular point there corresponds not one but two values of each of the parameters t and τ, namely, $t = \pm 1$ and $\tau = 0, \infty$ (see Figure 13).

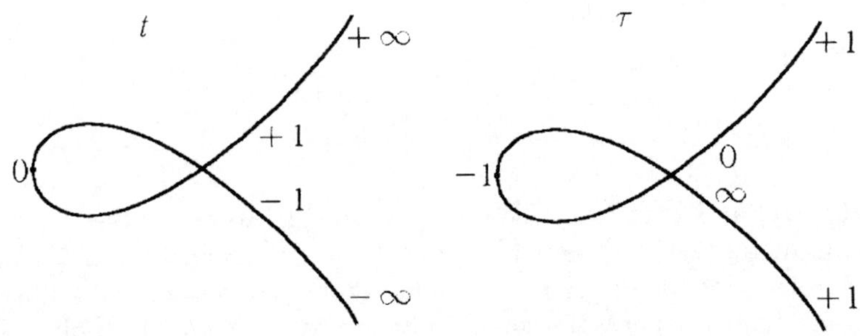

FIGURE 13

PROBLEMS

1.5.1. Prove that the points $(t_1^{-2}, t_1^{-3}), \ldots, (t_6^{-2}, t_6^{-3})$ on the curve $x^3 = y^2$ lie on a second degree curve if and only if $t_1 + \cdots + t_6 = 0$.

1.5.2. The curve $y^2 = x^2(x-1)$ has a rational parameterization $x = t^2 + 1$, $y = t^3 + t$. Prove that the points corresponding to the values of parameter t_1, t_2 and t_3 lie on one straight line if and only if $t_1t_2 + t_2t_3 + t_3t_1 = 1$.

§1.6. No nonsingular cubic admits a rational parameterization

The singular cubics that we studied in the preceding section admit a rational parameterization. Now we prove that none of the nonsingular cubics admits a rational parameterization. Recall that a nonsingular cubic can be reduced to the form $y^2 = x(x-1)(x-\lambda)$, where $\lambda \neq 0, 1$.

THEOREM. *If $\lambda \neq 0, 1$, then there are no polynomials P_1, P_2, Q_1, Q_2 such that the nonconstant functions $y(t) = P_1(t)/P_2(t)$ and $x(t) = Q_1(t)/Q_2(t)$ satisfy the relation $y^2 = x(x-1)(x-\lambda)$.*

PROOF. Suppose that $P_1(t)/P_2(t)$ and $Q_1(t)/Q_2(t)$ are not constants and

$$\frac{P_1^2}{P_2^2} = \frac{Q_1}{Q_2} \cdot \frac{Q_1 - Q_2}{Q_2} \cdot \frac{Q_1 - \lambda Q_2}{Q_2}.$$

Then we may assume that the polynomials P_1 and P_2 are relatively prime and so are Q_1 and Q_2. Since

$$P_1^2 Q_2^3 = P_2^2 Q_1(Q_1 - Q_2)(Q_1 - \lambda Q_2),$$

it follows that the polynomial P_2^2, which is relatively prime to P_1^2, is divisible by Q_2^3 and the polynomial Q_2^3, which is relatively prime to Q_1, $Q_1 - Q_2$, and $Q_1 - \lambda Q_2$, is divisible by P_2^2. Hence, the polynomials Q_2^3 and P_2^2 are proportional to each other. Therefore, by replacing P_1 with a proportional polynomial we can obtain the equality

$$P_1^2 = Q_1(Q_1 - Q_2)(Q_1 - \lambda Q_2). \tag{6.1}$$

Moreover, the polynomial Q_2^3 is the square of a polynomial; hence, Q_2 is also the square of a polynomial.

The polynomials Q_1, $Q_1 - Q_2$ and $Q_1 - \lambda Q_2$ are pairwise relatively prime and, therefore, the equality (6.1) implies that each of them is a perfect square. Thus, in the family of polynomials of the form $\alpha Q_1 + \beta Q_2$, where $\alpha, \beta \in \mathbb{C}$, there are 4 perfect squares, namely, Q_1, Q_2, $Q_1 - Q_2$ and $Q_1 - \lambda Q_2$; these polynomials are not proportional and they are distinct because $\lambda \neq 0, 1$.

To get a contradiction let us show that on the projective line $\alpha Q_1 + \beta Q_2$, where Q_1 and Q_2 are relatively prime, not more than three points can be perfect squares. Indeed, suppose that on this projective line there are four perfect squares:

$$R_1^2, \quad R_2^2, \quad \alpha_1 R_1^2 - \beta_1 R_2^2 \quad \text{and} \quad \alpha_2 R_1^2 - \beta_2 R_2^2.$$

Since the polynomials R_1 and R_2 are relatively prime, it follows that the polynomials $\sqrt{\alpha_i} R_1 \pm \sqrt{\beta_i} R_2$ should be perfect squares. As a result, from the projective line $\alpha Q_1 + \beta Q_2$ on which there are four perfect squares we come to the projective line $\alpha R_1 + \beta R_2$ on which there also are four perfect squares. From this projective line we can come to another projective line, etc. But each such passage decreases the maximal degree of every polynomial of the form $\alpha Q_1 + \beta Q_2$ at least by a factor of two. Contradiction. □

PROBLEMS

1.6.1. Give an example of relatively prime polynomials Q_1 and Q_2 for which the polynomials Q_1, $Q_1 + Q_2$ and $Q_1 + 2Q_2$ are perfect squares.

CHAPTER 2

Elliptic Functions

The addition of points on the circle is related to its parameterization by the functions *sine* and *cosine*. Indeed, consider the map $f : \mathbb{R} \longrightarrow S^1$ given by the formula $f(t) = (\cos t, \sin t)$. This map parameterizes the circle by real numbers in such a way that the addition of points on the circle corresponds to the addition of real numbers.

A similar parameterization exists for cubics. It is obtained by means of elliptic functions. Under this parameterization the addition of points on a cubic defined in Chapter 1 corresponds to the addition of the values of the parameter.

In this chapter we will study the main properties of elliptic functions and show how one can parameterize a nonsingular cubic with their help.

The name *elliptic functions* stems from *ellipse*, but the relation is rather indirect. It is elliptic integrals that are in direct relation to the ellipse. The length of an arc of the ellipse is expressed by an elliptic integral of a particular form. This is precisely where the name *elliptic integrals* stems from. Elliptic functions appeared in the process of inversion of elliptic integrals of another particular form not related to the computation of the arc length of an ellipse.

Elliptic integrals appeared as early as the seventeenth century in calculations of arc lengths of certain curves, primarily *ellipses*. Apart from the ellipse an interesting example is *Bernoulli's lemniscate* whose arc length is given by an integral of the form $\int_0^\alpha \frac{dx}{\sqrt{1-x^4}}$. It was for this integral that the Italian mathematician Count **Fagnano** obtained in the first half of the eighteenth century the addition theorem

$$\int_0^\alpha \frac{dx}{\sqrt{1-x^4}} + \int_0^\beta \frac{dx}{\sqrt{1-x^4}} = \int_0^\gamma \frac{dx}{\sqrt{1-x^4}},$$

where

$$\gamma = \frac{\alpha\sqrt{1-\beta^4} + \beta\sqrt{1-\alpha^4}}{1+\alpha^2\beta^2}.$$

Euler was interested in Fagnano's studies. Euler managed to obtain the addition theorem for integrals of an even more general form,

$$\int_0^\alpha \frac{dx}{\sqrt{P(x)}}, \quad \text{where} \quad P(x) = 1 + mx^2 + nx^4.$$

Namely, he proved that

$$\int_0^\alpha \frac{dx}{\sqrt{P(x)}} + \int_0^\beta \frac{dx}{\sqrt{P(x)}} = \int_0^\gamma \frac{dx}{\sqrt{P(x)}},$$

where

$$\gamma = \frac{\alpha\sqrt{P(\beta)} + \beta\sqrt{P(\alpha)}}{1 - n\alpha^2\beta^2}.$$

Euler also obtained addition theorems of an even more general form.

After Euler, **Legendre** tirelessly worked for many years on the development of the theory of elliptic integrals. He summarized the results of his studies in the book *Exercises de calcul intégral* (*Exercises on Integral Calculus*), published in 1811–1819. A revised edition of this book was issued in 1827–1832 under the name *Traité des fonctions elliptiques et des intégrales eulériennes* (*Treatise on Elliptic Functions and Euler Integrals*). These are three large volumes that contain a vast number of theorems on properties of elliptic integrals and their applications.

Legendre called *elliptic functions* what are nowadays called *elliptic integrals.* After works of **Abel** and **Jacobi** the importance of Legendre's book dwindled. Abel and Jacobi themselves, however, referred to Legendre's book with great respect, as befits it.

Elliptic function theory proper began with Abel's work *Recherches sur les fonctions elliptiques* (*Studies on Elliptic Functions*) published in 1827–1828 in Crelle's journal. Abel showed that the inversion of an elliptic integral of the first kind,

$$\alpha = \int \frac{dx}{\sqrt{(1 - cx^2)(1 + ex^2)}},$$

gives rise to a function $\varphi(\alpha)$ that has two periods in the complex domain. Abel meticulously studied the equations that relate $\varphi(\alpha)$ with $\varphi(n\alpha)$. Jacobi started to study elliptic function theory almost simultaneously with Abel. This led to a tense, albeit short, competition between them. Without a permanent position, almost in poverty, Abel finished the second part of the *Recherches* ... and continued his intensive studies. But soon he became seriously ill and died in 1829 at the age of 27.

Long before Abel and Jacobi, **Gauss** had known much of what was discovered by them. But Gauss did not publish his results.

* * *

It is convenient to consider elliptic functions as functions of a complex variable. Many of their properties are developed only on the complex plane[1] $\mathbb{C}$, and not on the real line $\mathbb{R}$. The parameterization of a cubic curve is also more graphic over $\mathbb{C}$. Therefore, we will start with the investigation of the topology of a nonsingular cubic in $\mathbb{C}P^2$. It turns out that from the topological point of view all such curves are alike: all of them are two-dimensional tori.

§2.1. The topological structure of nonsingular cubics in $\mathbb{C}P^2$

The equation of any nonsingular cubic in $\mathbb{C}P^2$ can be reduced to the form

$$y^2 z = (x - a_1 z)(x - a_2 z)(x - a_3 z), \tag{1.1}$$

[1]In algebraic geometry this set is usually referred to as *complex line*; but the authors use the term more usual among geometers. *Translator.*

where the numbers a_i are pairwise distinct. This equation determines a complex curve in $\mathbb{C}P^2$ whose complex dimension is equal to 1 and whose real dimension is equal to 2.

To find the topological structure of the curve (1.1) in $\mathbb{C}P^2$, consider the projection

$$p : \mathbb{C}P^2 \setminus \{(0,1,0)\} \longrightarrow \mathbb{C}P^1, \quad (x, y, z) \mapsto (x, z).$$

The complex projective line $\mathbb{C}P^1$ (a one-point compactification of $\mathbb{C}$) is homeomorphic to the two-dimensional sphere S^2. For $b \neq 0$ the equation $y^2 = b$ has precisely two distinct solutions. Therefore, if $z \neq 0$ and $x - a_i z \neq 0$, then a point $(x, z) \in \mathbb{C}P^1$ has exactly two preimages that belong to the curve (1.1). If $z \neq 0$ but x/z is equal to one of the numbers a_i, then there is only one preimage. For $z = 0$ the equation (1.1) turns into the equation $x^3 = 0$. Therefore, the point $\infty = (1, 0)$ also has only one preimage, namely, $(0, 1, 0)$. More exactly, the preimage of the point $(1, z)$ also tends to $(0, 1, 0)$ as $z \longrightarrow 0$.

The projection p of the curve (1.1) on $\mathbb{C}P^1$ is described as follows. If we exclude from $\mathbb{C}P^1$ the points a_1, a_2, a_3, and ∞, then all points have exactly two preimages. The structure of the map in vicinities of points a_i and ∞ should be studied in more detail. For simplicity, assume that $a_1 = 0$. Consider affine coordinates, i.e., set $z = 1$. The projection of the curve (1.1) on $\mathbb{C}P^1$ in this coordinate system can be described as $(x, y) \longmapsto x$. Then (1.1) takes the form

$$y^2 = x(x - a_2)(x - a_3),$$

where $a_2 a_3 \neq 0$. For points x close to zero the quantity $(x - a_2)(x - a_3)$ is almost a constant, i.e., we have almost the equation $y^2 = cx$. This equation has solutions of the form $x = c\lambda^2 e^{2i\varphi}, y = c\lambda e^{i\varphi}$. As φ varies from 0 to π, we perform a full revolution around the point (0, 1) on $\mathbb{C}P^1$. Under such a revolution, y changes sign. Lifting the revolution around (0, 1) on $\mathbb{C}P^1$ to the curve (1.1) we do not get back the initial point (see Figure 14). But by performing the revolution once again we return to the initial point, since by changing the sign of y_0 twice we get y_0.

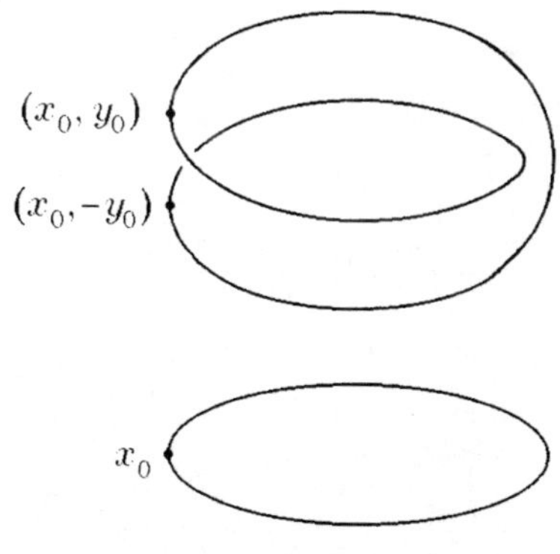

FIGURE 14

The structure of the projection of the curve (1.1) on $\mathbb{C}P^1$ in a vicinity of ∞ is the same as that in the vicinities of the a_i. Indeed, let $x = 1$. Then in a neighborhood of $z = 0$ the equation (1.1) looks approximately as $y^2 = \frac{1}{z}$ and, hence, the sign of y changes under a full revolution about the point $z = 0$.

Let us cut $\mathbb{C}P^1$ from a_1 to a_2 and from a_3 to ∞. The liftings of these cuts to the curve (1.1) divides it into two parts. Indeed, advancing along any closed path in $\mathbb{C}P^1$ that does not intersect the cuts we will circumvent the points a_1, a_2, a_3 and ∞ only in pairs and under the passage around two points the value of y does not change. Therefore, it is impossible to get from one preimage of a point of $\mathbb{C}P^1$ into its other preimage without intersecting the cuts.

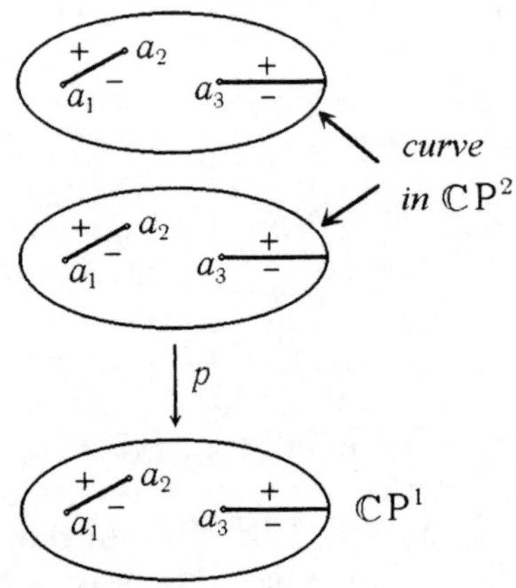

FIGURE 15

If we cut $\mathbb{C}P^1$ from a_1 to a_2 and from a_3 to ∞, then the remaining part of $\mathbb{C}P^1$ can be represented in the form of a plane with cuts, as in Figure 15. The part of the curve (1.1) that lies above this plane consists of two pieces. We only have to understand how to glue these pieces. Traversing a cut in $\mathbb{C}P^1$ we go from the boundary marked with a plus sign on one piece of the curve (1.1) to the boundary marked with a minus sign of the other piece. Hence, when the boundaries are glued, we get a torus (Figure 16).

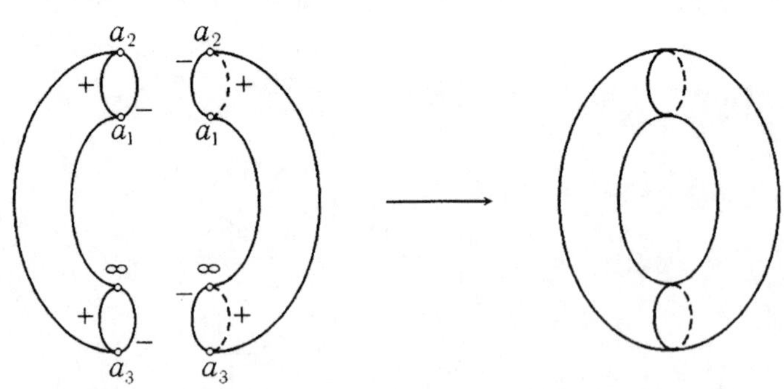

FIGURE 16

A parameterization of a cubic in $\mathbb{C}P^2$ can be determined by means of a map $f : \mathbb{C}^1 \longrightarrow \mathbb{C}P^2$, where $f(z) = (F_1(z), F_2(z), 1)$. The image of this map should be a torus. The simplest map of $\mathbb{C}^1$ to a torus is obtained by the identification of all points of the form $z + n\omega_1 + m\omega_2$. In other words, ω_1 and ω_2 are periods of the functions F_1 and F_2.

§2.2. The elliptic functions

A function f is called *doubly periodic* if $f(z + n\omega_1 + m\omega_2) = f(z)$ for any $m, n \in \mathbb{Z}$ and some ω_1 and ω_2 such that $\omega_1/\omega_2 \notin \mathbb{R}$; for definiteness, we assume that $\mathrm{Im}(\omega_1/\omega_2) > 0$. This means that the rotation from ω_1 to ω_2 on the complex plane[2] is performed clockwise (Figure 17).

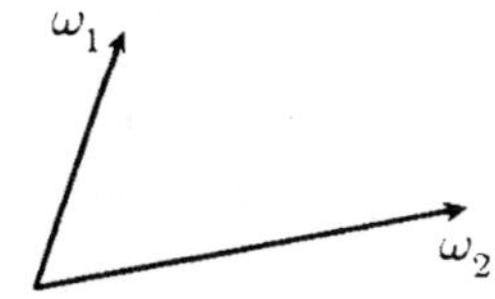

FIGURES 17

In what follows we will only be interested in meromorphic doubly periodic functions. Recall that an analytic function is called *meromorphic* if in the finite domain of $\mathbb{C}$ it has no singular points other than poles. In a vicinity of any finite point a a meromorphic function f can be expanded in the series

$$f(z) = c_0(z-a)^r + c_1(z-a)^{r+1} + \cdots,$$

where $c_0 \neq 0$ and r is an integer. A meromorphic doubly periodic function is called an *elliptic* function.

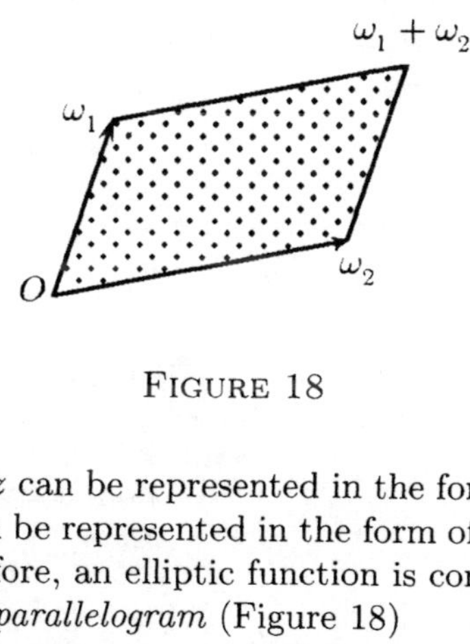

FIGURE 18

Any complex number z can be represented in the form $z = a_1\omega_1 + a_2\omega_2$, where $a_i \in \mathbb{R}$. The number a_i can be represented in the form of the sum of its integer and fractional parts and, therefore, an elliptic function is completely determined by its values in *the fundamental parallelogram* (Figure 18)

$$\{\alpha_1\omega_1 + \alpha_2\omega_2 \mid 0 \leq \alpha_1, \alpha_2 \leq 1\}.$$

The image of the fundamental parallelogram under any parallel translation will also be called the *fundamental parallelogram.*

2.2.1. THEOREM. *An elliptic function without poles is a constant.*

[2]By the complex plane we mean the real plane $\mathbb{R}^2$ with a complex structure, i.e., the complex plane is a 1-dimensional complex vector space.

PROOF. Suppose that an elliptic function $f(z)$ has no poles. Then the function $|f(z)|$ is continuous on $\mathbb{C}$. Since the fundamental parallelogram is compact, $|f(z)| \leq M$ for a number M. But then $|f(z)| \leq M$ for all $z \in \mathbb{C}$. Thus, f is a bounded analytic function on $\mathbb{C}$. By Liouville's theorem, f is a constant. □

All finite singular points of a meromorphic function are isolated. Hence, the fundamental parallelogram contains only a finite number of singular points. Therefore, there exists a parallel translation of the fundamental parallelogram such that there are no singular points on its sides. In what follows we will assume that there are no singular points on the sides of the fundamental parallelogram.

STATEMENT. *Let P be the fundamental parallelogram with the vertices $\alpha, \alpha + \omega_1, \alpha+\omega_1+\omega_2$ and $\alpha+\omega_2$, and ∂P its boundary (Figure 19). Then $\int_{\partial P} f(z)dz = 0$ for any elliptic function $f(z)$.*

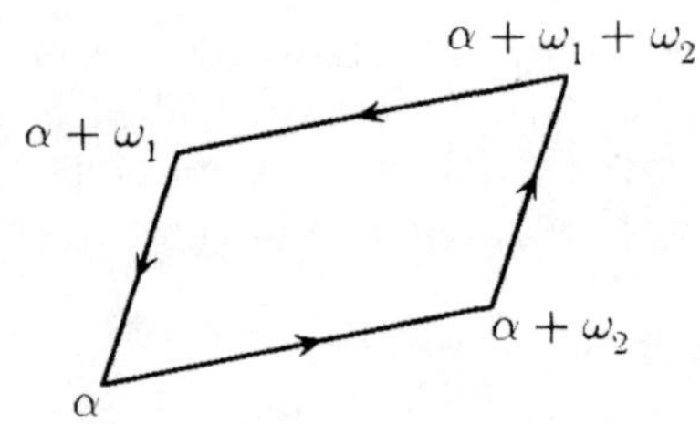

FIGURE 19

PROOF. Indeed, this integral is the sum of certain expressions among which there are, e.g., $\int_{\alpha}^{\alpha+\omega_1} f(z)dz$ and $\int_{\alpha+\omega_2}^{\alpha+\omega_1+\omega_2} f(z)dz$ with signs plus and minus, respectively. These integrals are equal, since $f(z+\omega_2) = f(z)$. Similarly, the integrals along the other pair of sides also cancel. □

This statement allows us to get essential information on zeros and poles of an elliptic function.

2.2.2. THEOREM. a) *The sum of the residues of an elliptic function $f(z)$ at singular points inside the fundamental parallelogram is equal to zero.*

b) *For an elliptic function, let a_i be those of its zeros and poles that lie inside the fundamental parallelogram, and r_i their orders (positive for zeros and negative for poles). Then $\sum r_i = 0$ and $\sum r_i a_i \equiv 0 \pmod{\Lambda}$, i.e., $\sum r_i a_i = m\omega_1 + n\omega_2$, where m and n are integers.*

PROOF. As we know from complex analysis, if on the boundary of the fundamental parallelogram P there are no singular points of a meromorphic function g, then

$$\sum_P \operatorname{res} g = \frac{1}{2\pi i} \int_{\partial P} g(z)dz.$$

To prove part a) it suffices to use this identity for $g = f$.

To prove that $\sum r_i = 0$ and $\sum r_i a_i \equiv 0 \pmod{\Lambda}$ we set $g(z) = f'(z)/f(z)$ and $g(z) = zf'(z)/f(z)$, respectively.

If $f(z)$ is an elliptic function, then the function $g(z) = f'(z)/f(z)$ is also an elliptic function. Moreover, if $f(z) = c_0(z-a)^r + c_1(z-a)^{r+1} + \cdots$, then $g(z) =$

$r(z-a)^{-1}+d_1+d_2(z-a)+\cdots$ and, therefore, the residue of $g(z)$ at the point a is equal to r. Thus, $\sum r_i = 0$.

To prove the identity $\sum r_i a_i = m\omega_1 + n\omega_2$ we must perform certain calculations, since the function $g(z) = zf'(z)/f(z)$ is not necessarily an elliptic one. First, observe that if $f(z) = c_0(z-a)^r + c_1(z-a)^{r+1}+\cdots$, then

$$g(z) = \frac{a+(z-a)}{z-a}\cdot\frac{rc_0(z-a)^{r-1}+\cdots}{c_0(z-a)^{r-1}+\cdots} = ar(z-a)^{-1}+\cdots;$$

hence, the residue of $g(z)$ at the point a is equal to ar. Now, let us compute the integral $\int_{\partial P} g(z)dz$. The difference of the integrals

$$\int_\alpha^{\alpha+\omega_1}\frac{zf'(z)}{f(z)}dz$$

and

$$\int_{\alpha+\omega_2}^{\alpha+\omega_1+\omega_2}\frac{zf'(z)}{f(z)}dz = \int_\alpha^{\alpha+\omega_1}\frac{(z+\omega_2)f'(z)}{f(z)}dz$$

contributes to this integral. This difference is equal to

$$-\omega_2\int_\alpha^{\alpha+\omega_1}\frac{f'(z)}{f(z)}dz = -\omega_2 \ln f(z)|_\alpha^{\alpha+\omega_1}.$$

Since $f(\alpha+\omega_1) = f(\alpha)$, the logarithm of $f(z)$ can change only by $2k\pi i$ as z varies from α to $\alpha+\omega_1$. As a result we see that contribution to the integral

$$\frac{1}{2\pi i}\int_{\partial P} g(z)dz$$

from one pair of sides of the parallelogram is $n\omega_2$, where n is an integer. Similarly, the contribution from the other pair is $m\omega_1$. □

As we have already said, a nonconstant elliptic function must have at least one pole inside the fundamental parallelogram. But since the sum of the residues at the singular points that lie inside the fundamental parallelogram is equal to zero, the function cannot have exactly one pole of order 1 there. For an elliptic function, the number (multiplicities counted) of poles inside the fundamental parallelogram is called the *order* of the elliptic function. The minimal possible order is thus equal to 2; and there are two ways for such a possibility to realize itself:

1) one pole of order 2, i.e., of multiplicity 2 (this takes place for the *Weierstrass function* that will be discussed in the next section);

2) two simple poles (this takes place for the *Jacobi elliptic functions* that will be discussed in §2.8).

By Theorem 2.2.2 b), for an elliptic function the sum of orders of the zeros inside the fundamental parallelogram is equal to the sum of orders of the poles, i.e., is equal to the order of the function. It is also clear that the poles of the function $f(z)-c$ are the same as those of $f(z)$. Therefore, an elliptic function of order r takes any finite value inside the fundamental parallelogram exactly r times (multiplicities counted).

PROBLEMS

2.2.1. The elliptic functions f and g have the same periods and at every pole they have the same principal parts

$$c_r(z-a)^r + c_{r+1}(z-a)^{r+1} + \cdots + c_{-1}(z-a)^{-1} \qquad (\text{here } r<0).$$

Prove that the difference of these functions is a constant.

2.2.2. The elliptic functions f and g have the same periods and their zeros and poles (multiplicities counted) coincide. Prove that the ratio of these functions is a constant.

§2.3. The Weierstrass function

We have already proved certain properties of elliptic functions but we have not yet established that there exist nonconstant elliptic functions. It is time to give an example of a nontrivial elliptic function. Let us show that for any lattice Λ the function

$$\wp(z) = \frac{1}{z^2} + \sum{}' \left[\frac{1}{(z-\omega)^2} - \frac{1}{\omega^2}\right], \tag{3.1}$$

where the sum runs over all nonzero elements $\omega \in \Lambda$, is an elliptic function. (The fact that the summation runs over nonzero elements only is denoted by a prime.) The grouping of terms in square brackets is essential, since each of the series $\sum'(z-\omega)^{-2}$ and $\sum'\omega^{-2}$ diverges.

First, let us prove that the series (3.1) does indeed define a meromorphic function. On any compact set K that does not contain the points of the lattice, this series converges uniformly and absolutely. Indeed,

$$\frac{1}{(z-\omega)^2} - \frac{1}{\omega^2} = \frac{2z\omega - z^2}{\omega^2(z-\omega)^2} = \frac{\omega}{\omega^4}\cdot\frac{2z - z^2\omega^{-1}}{(z\omega^{-1}-1)^2}.$$

If $|\omega|$ is sufficiently large, then $\frac{2z-z^2\omega^{-1}}{(z\omega^{-1}-1)^2} \approx 2z$. Therefore, for all $\omega \in \Lambda'$ with a sufficiently large value of $|\omega|$ and for all $z \in K$ there exists a constant C such that

$$\left|\frac{1}{(z-\omega)^2} - \frac{1}{\omega^2}\right| < \frac{C}{|\omega|^3}.$$

Moreover, $|z-\omega| > \varepsilon$ for all $z \in K$ and $\omega \in \Lambda'$; hence, such a constant exists for all $\omega \in \Lambda'$ as well. It is easy to verify that the series $\sum' |\omega|^{-3}$ converges. Indeed,

$$\sum{}' |\omega|^{-3} = \sum_{n=1}^{\infty} \sum_{\max(p,q)=n} |p\omega_1 + q\omega_2|^{-3} \le \sum_{n=1}^{\infty} 8n(nh)^{-3},$$

where $h = \min(|\omega_1|, |\omega_2|)$ is the smallest of two sides of the fundamental parallelogram. Thus, $\wp(z)$ is a meromorphic function with poles at the nodes of the lattice. It is called *the Weierstrass function.* Let us prove the periodicity of $\wp(z)$. For that, let us consider its derivative

$$\wp'(z) = -2\sum (z-\omega)^{-3}.$$

Here the summation runs over all the nodes of the lattice. Clearly, ω_1 and ω_2 are periods of the function $\wp'(z)$. Hence, the functions $\wp(z+\omega_i)$ and $\wp(z)$ can only differ by a constant c. Substituting $z = -\omega_i/2$ into the equality $\wp(z+\omega_i) = \wp(z) + c$

we get $\wp(\omega_i/2) = \wp(-\omega_i/2) + c$. But from formula (3.1) it is clear that the function $\wp(z)$ is even. Hence, $c = 0$, i.e., ω_1 and ω_2 are periods of $\wp(z)$.

The function $\wp$ has double poles at the nodes of the lattice; and it has no other singular points. Inside the fundamental parallelogram there is exactly one node of the lattice. Therefore, inside the fundamental parallelogram the sum of the poles of $\wp$ is congruent to zero modulo Λ. By Theorem 2.2.2, inside the fundamental parallelogram, there are two zeros of $\wp$, call them u and v, such that $u + v \equiv 0 \pmod{\Lambda}$. For any constant c the poles of the function $\wp(z) - c$ coincide with the poles of the function $\wp(z)$ and, therefore, inside the fundamental parallelogram there are exactly two points, u and v, for which $\wp(u) = \wp(v) = c$ and $u+v \equiv 0 \pmod{\Lambda}$. If $u \equiv -u \pmod{\Lambda}$, then these two points coincide, i.e., the corresponding value of $\wp$ is attained twice. At the points where the two zeros of $\wp(z) - c$ merge the derivative $\wp'(z)$ vanishes. It is possible to select the fundamental parallelogram so that it contains exactly four points for which $u \equiv -u \pmod{\Lambda}$, namely, the points

$$0, \quad \frac{\omega_1}{2}, \quad \frac{\omega_2}{2} \quad \text{and} \quad \frac{\omega_1 + \omega_2}{2}$$

(Figure 20). The first of these points is the pole of $\wp$ and the other three points are zeros of $\wp'$.

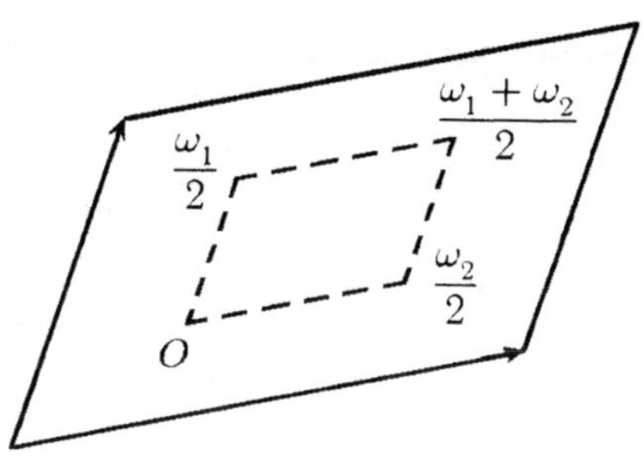

FIGURE 20

Thus, the values

$$e_1 = \wp\left(\frac{\omega_1}{2}\right), \quad e_2 = \wp\left(\frac{\omega_1 + \omega_2}{2}\right) \quad \text{and} \quad e_3 = \wp\left(\frac{\omega_2}{2}\right)$$

of $\wp$ are of multiplicity 2 and there are no other values of multiplicity 2. The values of multiplicity 2 correspond to the zeros of the derivative; hence, $\wp'(z) = 0$ if and only if

$$z \equiv \frac{1}{2}\omega_1, \quad \frac{1}{2}(\omega_1 + \omega_2), \quad \frac{1}{2}\omega_2 \pmod{\Lambda}.$$

Observe that the numbers e_1, e_2 and e_3 are distinct. Suppose, for example, that $e_1 = e_3$. Then the function $\wp(z) - e_1$ has zeros of multiplicity 2 at the points $\omega_1/2$ and $\omega_2/2$, i.e., inside the fundamental parallelogram there are at least 4 zeros of this function. This is impossible.

The Weierstrass function not only gives an example of an elliptic function but enables one to describe the structure of all elliptic functions.

THEOREM. *Let $f(z)$ be an arbitrary elliptic function and $\wp(z)$ the Weierstrass function with the same periods. Then there exist rational functions R and R_1 such that*

$$f = R(\wp) + R_1(\wp)\wp'.$$

PROOF. It is possible to represent $f(z)$ as the sum of the even function $g(z) = \frac{1}{2}(f(z) + f(-z))$ and the odd function $h(z) = \frac{1}{2}(f(z) - f(-z))$. Since $\wp'(z)$ is an odd function, $h_1(z) = h(z)/\wp'(z)$ is an even function and for the even functions g and h_1 we have

$$f(z) = g(z) + h_1(z)\wp'(z).$$

Therefore, it suffices to prove that any even elliptic function can be represented as a rational function of $\wp$.

First, let us prove certain properties of zeros and poles of an even elliptic function.

1°. Let f be an even function, and u its zero (resp. pole) of order m. Then $-u$ is also a zero (resp. pole) of order m. Indeed, in the case of zeros it suffices to observe that for an even function f we have

$$f^{(k)}(-z) = (-1)^k f^{(k)}(z).$$

In the case of poles we may consider $\frac{1}{f}$ instead of f.

2°. If f is an even elliptic function and $u \equiv -u \pmod{\Lambda}$, then the order of a zero or pole of f at u is even. We will prove this for zeros (since for poles we may consider $\frac{1}{f}$ instead of f). The condition $u \equiv -u \pmod{\Lambda}$ is equivalent to the fact that

$$u \equiv 0,\ \frac{\omega_1}{2},\ \frac{\omega_1+\omega_2}{2},\ \frac{\omega_2}{2} \pmod{\Lambda}.$$

Moreover, the periodicity of f' implies that $f'(u) = f'(-u)$. But the derivative of an even function is odd; hence, $f'(u) = 0$. Therefore, if the function f has a zero at u, then this zero is of multiplicity at least two. For any of the cases

$$u \equiv \frac{\omega_1}{2},\ \frac{\omega_1+\omega_2}{2},\ \frac{\omega_2}{2} \pmod{\Lambda}$$

the function $F(z) = \wp(z) - \wp(u)$ has a zero of order 2 at u and if $u \equiv 0 \pmod{\Lambda}$, then the function $F(z) = 1/\wp(z)$ has such a property. Using F, we can construct an even elliptic function $f_1(z) = f(z)/F(z)$ for which the order of the zero at u is less by 2 than that of f. Hence, if $f_1(u) \neq 0$, then the order of the zero of f at u is equal to 2 and if $f_1(u) = 0$, then we can apply the same arguments to f_1 instead of f, etc.

By the above properties of zeros and poles of the even elliptic function f they can be divided into pairs of the form $(x, -x)$. Select a representative from each such pair with $a_1, \dots, a_k$ the representatives of the zeros and $b_1, \dots, b_k$ the representatives of the poles. Consider the elliptic function

$$Q(z) = R(\wp(z)) = \frac{\prod(\wp(z) - \wp(a_i))}{\prod(\wp(z) - \wp(b_i))},$$

where we take only those a_i and b_i that are distinct from the nodes of the lattice (since at the nodes the function $\wp$ takes infinite values). If we disregard the nodes of the lattice, then the complete system of zeros and poles of Q is the same as that of f, since $\wp(z) = \wp(a)$ if and only if $z \equiv \pm a \pmod{\Lambda}$. But by Theorem 2.2.2 b) for an elliptic function the sum of orders of its zeros and poles inside the fundamental parallelogram is equal to zero; hence, the order of a zero or a pole at a node of the lattice is uniquely determined by the orders of the other zeros and poles. Therefore, $f(z)/Q(z)$ is an elliptic function without poles, i.e., a constant. As a result, we see that $f(z) = cR(\wp(z))$. □

PROBLEMS

2.3.1. All the poles of an elliptic function f lie at the nodes of the period lattice. Prove that $f = P(\wp) + P_1(\wp)\wp'$, where P and P_1 are polynomials.

HINT. From the decomposition $f = R(\wp) + R_1(\wp)\wp'$ we can derive that

$$2R(\wp(z)) = f(z) + f(-z) \quad\text{and}\quad 2\wp'(z)R_1(\wp(z)) = f(z) - f(-z).$$

Prove that the functions R and R_1 do not take infinite values at finite points.

§2.4. A differential equation for the Weierstrass function $\wp(z)$

In the preceding section we proved that an even elliptic function can be rationally expressed in terms of $\wp(z)$ and the expression was explicitly described. This can be applied to the even function $(\wp'(z))^2$. It has zeros of multiplicity 2 at $\frac{\omega_1}{2}$, $\frac{\omega_1+\omega_2}{2}$, $\frac{\omega_2}{2}$ and a pole of multiplicity 6 at a node of the lattice. Hence,

$$(\wp'(z))^2 = c(\wp(z) - e_1)(\wp(z) - e_2)(\wp(z) - e_3), \tag{4.1}$$

where $e_1 = \wp(\frac{\omega_1}{2}), e_2 = \wp(\frac{\omega_1+\omega_2}{2})$ and $e_3 = \wp(\frac{\omega_2}{2})$. Since $\wp(z) = z^{-2} + \cdots$ and $\wp'(z) = -2z^{-3} + \cdots$, it follows that $c = 4$.

There is also another way to obtain a differential equation for $\wp(z)$. It not only gives us a new method to deduce this equation, but also provides another form of this equation. We will use the fact that if the coefficients of nonpositive powers of z in the Laurent expansions of the functions $(\wp'(z))^2$ and $a\wp^3(z)+b\wp^2(z)+c\wp(z)+d$ coincide, then these functions are equal. Indeed, their difference is an elliptic function without poles and at 0 its value is equal to 0. Hence, their difference is a constant equal to 0.

Since

$$\left(\frac{1}{1-x}\right)^2 = \frac{d}{dx}\left(\frac{1}{1-x}\right) = 1 + 2x + 3x^2 + \cdots,$$

it follows that

$$\begin{aligned}\wp(z) &= \frac{1}{z^2} + \sum{}'\left(\frac{1}{(z-\omega)^2} - \frac{1}{\omega^2}\right)\\ &= \frac{1}{z^2} + \sum{}'\left(\frac{1}{\omega^2}\left(1 + 2\frac{z}{\omega} + 3\left(\frac{z}{\omega}\right)^2 + \cdots\right) - \frac{1}{\omega^2}\right)\\ &= \frac{1}{z^2} + 3G_4z^2 + 5G_6z^4 + \cdots,\end{aligned}$$

where $G_k = \sum{}'\omega^{-k}$ (for k odd this sum is equal to zero). Hence,

$$\begin{aligned}\wp(z) = z^{-2} + \cdots, \wp^2(z) &= z^{-4} + 6G_4 + \cdots,\\ \wp^3(z) &= z^{-6} + 9G_4z^{-2} + 15G_6 + \cdots,\\ (\wp'(z))^2 &= 4z^{-6} - 24G_4z^{-2} - 80G_6 + \cdots\end{aligned}$$

(only the terms of the Laurent expansion of interest to us are written). Thus,

$$a\wp^3(z)+b\wp^2(z)+c\wp(z)+d = az^{-6}+bz^{-4}+(9aG_4+c)z^{-2}+(15aG_6+6bG_4+d)+\cdots.$$

Therefore, $a\wp^3 + b\wp^2 + c\wp + d = (\wp')^2$ if

$$a = 4, \quad b = 0, \quad 9aG_4 + c = -24G_4 \quad \text{and} \quad 15aG_6 + 6bG_4 + d = -80G_6,$$

i.e.,

$$a = 4, \quad b = 0, \quad c = -60G_4 \quad \text{and} \quad d = -140G_6.$$

Set $g_2 = 60G_4 = 60\sum{}'\omega^{-4}$ and $g_3 = 140G_6 = 140\sum{}'\omega^{-6}$. Then

$$(\wp'(z))^2 = 4\wp^3(z) - g_2\wp(z) - g_3. \tag{4.2}$$

Comparing (4.1) with (4.2) we see that

$$e_1 + e_2 + e_3 = 0, \quad e_1e_2 + e_2e_3 + e_3e_1 = -\frac{g_2}{4} \quad \text{and} \quad e_1e_2e_3 = -\frac{g_3}{4}.$$

It is easy to verify that

$$g_2^3 - 27g_3^2 = 16(e_1 - e_2)^2(e_2 - e_3)^2(e_3 - e_1)^2.$$

In the preceding section it was shown that the numbers e_1, e_2 and e_3 are distinct. Therefore, $g_2^3 - 27g_3^2 \neq 0$. A natural question arises:

Given numbers g_2 and g_3 such that $g_2^3 \neq 27g_3^2$, is there a lattice for which $g_2 = 60\sum{}'\omega^{-4}$ and $g_3 = 140\sum{}'\omega^{-6}$?

The answer to this question is an affirmative one; see, e.g., [**B9**].

Problems

2.4.1. Prove that $\wp'' = 6\wp^2 - g_2/2$ and $\wp''' = 12\wp'\wp$.

Hint. Differentiate (4.2).

2.4.2. Prove that $\wp^{(2n-2)}(z)$ and $\wp^{(2n+1)}(z)/\wp'(z)$ are polynomials in $\wp(z)$ of degree n.

2.4.3. Prove that $e_1^2 + e_2^2 + e_3^2 = \frac{g_2}{2}$, $e_1^3 + e_2^3 + e_3^3 = \frac{3g_3}{4}$ and $e_1^4 + e_2^4 + e_3^4 = \frac{g_2^2}{8}$.

Hint. Use the fact that $a + b + c$ divides both $a^3 + b^3 + c^3 - 3abc$ and $(a^2 + b^2 + c^2)^2 - 2(a^4 + b^4 + c^4)$.

§2.5. A parameterization of the cubic with the help of the Weierstrass function

The differential equation for $\wp$ enables us to clarify the nature of the addition of points on the cubic. To this end we have to use the fact that we left the following without proof earlier:

For any numbers g_2 and g_3 such that $g_2^3 \neq 27g_3^2$ there exists a lattice for which the Weierstrass function satisfies the equation

$$(\wp')^2 = 4\wp^3 - g_2\wp - g_3.$$

The cubic curve $y^2 = 4x^3 - g_2x - g_3$ can be parameterized with the help of $\wp$ setting $x = \wp(z)$ and $y = \wp'(z)$. Passing to homogeneous coordinates in $\mathbb{C}P^2$ the map $f : \mathbb{C}/\Lambda \longrightarrow \mathbb{C}P^2$ can be defined as

$$z \longmapsto \begin{cases} (\wp(z), \wp'(z), 1) & \text{for } z \neq 0, \\ (z^3\wp(z), z^3\wp'(z), z^3) = (0,1,0) & \text{for } z = 0. \end{cases}$$

Obviously, this map is analytic at all points distinct from the nodes of the lattice. Expressing it in the form

$$z \longmapsto \left(\frac{\wp(z)}{\wp'(z)}, 1, \frac{1}{\wp'(z)}\right)$$

we can verify that it is analytic in a neighborhood of the node of the lattice as well. The map f is a one-to-one map of the torus $\mathbb{C}/\Lambda$ to the cubic $y^2z = 4x^3 - g_2xz^2 - g_3z^3$ in $\mathbb{C}P^2$.

Indeed, on the infinite line $z = 0$ there lies only the point $(0,1,0)$ of this curve; the nodes of the lattice which correspond to one point on the torus are mapped into this point. For all other points we can consider an affine curve $y^2 = 4x^3 - g_2x - g_3$ and the map $z \longmapsto (\wp(z), \wp'(z))$.

The equation $\wp(z) = c$ may have either one or two solutions. It has two solutions if $\wp'(z) \neq 0$. The solutions then are of the form $\pm z$. The images of these two points under the map $z \longmapsto (\wp(z), \wp'(z))$ do not coincide, since the nonzero numbers $\wp'(z)$ and $\wp'(-z) = -\wp'(z)$ differ by a sign.

The addition of complex numbers induces an addition of the points on a torus which, in turn, with the help of the map f induces an addition of the points on a cubic. It turns out that this is precisely the addition of points on a cubic defined in §1.1 if for the zero element we take the infinite point (0, 1, 0).

Let the points P_1 and P_2 on a cubic correspond to the points z_1 and z_2 in $\mathbb{C}$, i.e., $P_i = (\wp(z_i), \wp'(z_i))$. Let us draw the straight line $y = ax + b$ through P_1 and P_2. Then $\wp'(z_i) = a\wp(z_i) + b$, where $i = 1, 2$.

At the point $z = 0$ the elliptic function $\wp'(z) - a\wp(z) - b$ has a pole of multiplicity 3 and it has no other poles at the other points of the fundamental parallelogram. Therefore, the order of this function is equal to 3, i.e., it has precisely 3 zeros, namely, the already known zeros z_1 and z_2 and a third zero z_3. Since the sum of the poles and zeros is equal to zero, we have $z_1 + z_2 + z_3 \equiv 0 \pmod{\Lambda}$, i.e., $z_3 \equiv -z_1 - z_2 \pmod{\Lambda}$. Thus, the third point of intersection of the line P_1P_2 with the cubic is the point

$$\begin{aligned} P_3' &= (\wp(z_3), \wp'(z_3)) = (\wp(-z_1 - z_2), \wp'(-z_1 - z_2)) \\ &= (\wp(z_1 + z_2), -\wp(z_1 + z_2)). \end{aligned}$$

Therefore, the point $P_3 = (\wp(z_1 + z_2), \wp'(z_1 + z_2))$ corresponding to the sum of z_1 and z_2 is symmetric to P_3' with respect to the x-axis (see Figure 21 on the next page). In other words, P_3 is the point of intersection of the cubic with the straight line $P_3'E$, where $E = (0,1,0)$ is the infinite point on the cubic. This is precisely what we wanted to establish.

By extending the above arguments a bit further, one can show that there exists an algebraic addition theorem for $\wp$, i.e., $\wp(z_1 + z_2)$ can be algebraically expressed in terms of $\wp(z_1)$ and $\wp(z_2)$. Indeed, the line $y = ax + b$ passing through the points

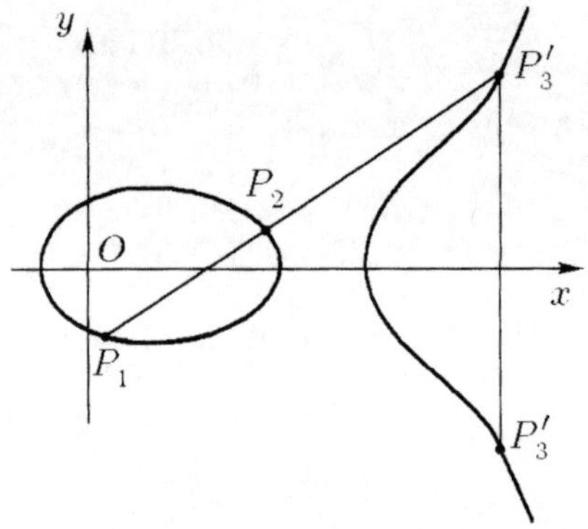

FIGURE 21

P_1 and P_2 intersects the cubic $y^2 = 4x^3 - g_2x - g_3$ at the three points (x_i, y_i), where $x_1 = \wp(z_1), x_2 = \wp(z_2)$ and $x_3 = \wp(z_1 + z_2)$. Therefore, the cubic equation

$$(ax + b)^2 = 4x^3 - g_2x - g_3$$

has the indicated roots x_1, x_2 and x_3. Expressing the coefficient of x^2 in terms of these roots we get

$$\wp(z_1) + \wp(z_2) + \wp(z_1 + z_2) = \frac{a^2}{4}.$$

Since $\wp'(z_1) = a\wp(z_1) + b$ and $\wp'(z_2) = a\wp(z_2) + b$, it follows that $a = \frac{\wp'(z_1)-\wp'(z_2)}{\wp(z_1)-\wp(z_2)}$. Therefore,

$$\wp(z_1 + z_2) = -\wp(z_1) - \wp(z_2) + \frac{1}{4}\left(\frac{\wp'(z_1) - \wp'(z_2)}{\wp(z_1) - \wp(z_2)}\right)^2.$$

Thus, $\wp(z_1 + z_2)$ can be rationally expressed in terms of the $\wp(z_i)$ and $\wp'(z_i)$, $i = 1, 2$. It remains to recall that $\wp'(z_i)$ can be algebraically expressed in terms of $\wp(z_i)$, namely:

$$\wp'(z_i) = \sqrt{4\wp^3(z_i) - g_2\wp(z_i) - g_3}.$$

* * *

With the help of the Weierstrass function one can also parameterize the curve

$$y^2 = G_4(x),$$

where $G_4(x) = a_0x^4 + a_1x^3 + a_2x^2 + a_3x + a_4$ is a fourth degree polynomial without multiple roots. To this end let us make the change of variables $x = x_1^{-1} + \alpha$, $y = y_1x_1^{-2}$. We get

$$y_1^2x_1^{-4} = b_4x_1^{-4} + b_3x_1^{-3} + b_2x_1^{-2} + b_1x_1^{-1} + G_4(\alpha).$$

If α is a root of G_4, then $y_1^2 = b_1x_1^3 + b_2x_1^2 + b_3x_1 + b_4$. This cubic can be parameterized with the help of the Weierstrass function. Notice that the change of variables

$$x = x_1^{-1} + \alpha, \quad y = y_1x_1^{-n}$$

enables us to pass in a similar way from the curve $y^2 = G_{2n}(x)$ to the curve $y^2 = G_{2n-1}(x)$.

§2.6. The elliptic integrals

The Weierstrass function $\wp(z)$ satisfies, as we have seen, the differential equation

$$\left(\frac{d\wp}{dz}\right)^2 = 4\wp^3 - g_2\wp - g_3.$$

Therefore,

$$dz = \frac{d\wp}{\sqrt{4\wp^3 - g_2\wp - g_3}},$$

i.e.,

$$z = \int \frac{du}{\sqrt{4u^3 - g_2u - g_3}},$$

where $u = \wp(z)$. Thus, $z = \wp^{-1}(u)$, i.e., the inversion of the integral

$$\int \frac{du}{\sqrt{4u^3 - g_2u - g_3}}$$

gives rise to the Weierstrass function.

An *elliptic integral* is an integral of the form

$$\int R(x, \sqrt{G(x)})dx,$$

where $G(x)$ is a polynomial of degree 3 or 4 without multiple roots and $R(x, y)$ is a rational function of two variables. At first, such integrals appeared in the calculations of arc lengths of various curves, for instance, ellipses. Only later it was noticed that for certain elliptic integrals the inverse functions possess more interesting properties, primarily, double periodicity.

Elliptic integrals can be reduced to certain simpler integrals. Let us preliminarily consider integrals of a form less involved than that of elliptic integrals. First of all, let us prove that if $R(x)$ is a rational function, then $\int R(x)dx$ is the sum of a rational function and a certain number of summands of the form $c_i \ln(x - a_i)$. It suffices to prove that a rational function $R(x)$ can be represented in the form

$$A(x) + \sum_{i,k} \frac{c_{i,k}}{(x - a_i)^k},$$

where $A(x)$ is a polynomial. Let $R(x) = P(x)/Q(x)$, where P and Q are polynomials. Dividing P by Q with a remainder we can pass to the fraction S/Q, where $\deg S < \deg Q$. Let $Q = Q_1Q_2$, where Q_1 and Q_2 are relatively prime polynomials. Then there exist polynomials a and b such that $a(x)Q_1(x) + b(x)Q_2(x) = 1$. Therefore,

$$\frac{S}{Q_1Q_2} = \frac{a_1SQ_1 + a_2SQ_2}{Q_1Q_2} = \frac{a_1S}{Q_2} + \frac{a_2S}{Q_1}.$$

In the fractions obtained one should also divide the numerator by the denominator with a remainder. After several such operations we arrive at the sum of a polynomial $A(x)$ and several fractions of the form $p(x)(x-a)^{-n}$, where $\deg p(x) < n$. The proof is completed by expressing the polynomial $p(x)$ in the form

$$p(x) = b_1(x - a)^{n-1} + b_2(x - a)^{n-2} + \cdots + b_n.$$

Leibniz was the first to study integration of rational functions. He only considered factorization of polynomials into factors with real coefficients and, therefore, he faced the question: *whether or not it is true that any real polynomial can be factored into factors of degree* 1 *and* 2 *with real coefficients.* In 1702 Leibniz published a paper in which he claimed that it is impossible to factor the polynomial $x^4 + a^4$ in the required fashion since

$$\begin{aligned} x^4 + a^4 &= (x^2 + a^2\sqrt{-1})(x^2 - a^2\sqrt{-1}) \\ &= (x + a\sqrt{\sqrt{-1}})(x - a\sqrt{\sqrt{-1}})(x + a\sqrt{-\sqrt{-1}})(x - a\sqrt{-\sqrt{-1}}) \end{aligned}$$

and the product of any two of these factors cannot be, as he believed, a quadratic with real coefficients. Only 17 years after that **Nicholas Bernoulli** (1687–1759) indicated that

$$x^4 + a^4 = (x^2 + a^2)^2 - 2a^2x^2 = (x^2 + \sqrt{2}ax + a^2)(x^2 - \sqrt{2}ax + a^2).$$

In their correspondence Leibniz and Jacob Bernoulli also discussed integrals of irrational expressions that appear in the study of various physical and mathematical problems. Many of these integrals are elliptic ones.

Let us now pass from rational functions to the simplest irrationalities. To calculate the integral

$$\int R(x, \sqrt{G(x)})dx,$$

where $G(x) = ax + b$ is a linear function, we first make the change of variables $u = ax + b$. As a result we get an integral of the form

$$\int R_1(u, \sqrt{u})du,$$

where R_1 is again a rational function. Now, set $t = \sqrt{u}$. Then $du = d(t^2) = 2tdt$ and, therefore,

$$\int R_1(u, \sqrt{u})du = \int R_1(t^2, t)2tdt = \int R_2(t)dt,$$

where R_2 is a rational function.

Now, let $G(x) = ax^2 + bx + c$. As has been said in the preceding section, with the help of the change of variables $x = x_1^{-1} + \alpha, y = y_1x_1^{-1}$ we can pass from the curve $y^2 = G(x)$ to the curve $y_1^2 = G_1(x_1)$, where G_1 is a linear function. Let us apply this change of variables in order to calculate the integral $\int R(x, y)dx$, where $y^2 = G(x)$. Let $x = x_1^{-1} + \alpha$ and $y = y_1x_1^{-1}$, where $G(\alpha) = 0$. Then $dx = -x_1^{-2}dx_1$ and

$$\int R(x, y)dx = -\int R(x_1^{-1} + \alpha, y_1x_1^{-1})x_1^{-2}dx_1 = \int R_1(x_1, y_1)dx_1,$$

where $y_1^2 = Ax_1 + B$.

Thus, the integrals of the form $\int R(x, y)dx$, where R is a rational function and $y = \sqrt{G(x)}$, can be expressed in terms of elementary functions if $\deg G \leq 2$. In the case when $\deg G = 3$ there may appear functions which are inverse to elliptic ones. The integral $\int R(x, y)dx$, where $y = \sqrt{G_4(x)}$, reduces to the integral $\int Q(x, y)dx$, where $y = \sqrt{4x^3 - g_2x - g_3}$. Indeed, using the change of variables $x = x_1^{-1} + \alpha$, $y = y_1x_1^{-2}$ we can pass from the fourth degree polynomial G_4 to a third degree polynomial and from an arbitrary third degree polynomial we can pass with the help of a linear change to a polynomial of the form $4x^3 - g_2x - g_3$.

We could have confined ourselves to the calculation of integrals of the form $\int R(x,y)dx$, where $y^2 = 4x^3 - g_2x - g_3$, but in many cases certain other forms of elliptic integrals are convenient. Therefore, we will first calculate elliptic integrals in the general form and later on we will study certain special forms.

Let $I = \int R(x,y)dx$, where R is a rational function, and

$$y^2 = a_0x^4 + 4a_1x^3 + 6a_2x^2 + 4a_3x + a_4,$$

where at least one of the coefficients a_0 and a_1 is nonzero.

THEOREM (Legendre). *The elliptic integral I can be represented as a linear combination of a rational function in x and y, the integral of a rational function of x, and of the integrals*

$$\int \frac{dx}{y}, \int \frac{xdx}{y}, \int \frac{x^2dx}{y} \quad \textit{and} \quad \int \frac{dx}{(x-c)y}.$$

PROOF. Since y^2 can be polynomially expressed in terms of x, we can assume that a rational function R does not contain y^k for $k \geq 2$. Moreover,

$$\frac{a+by}{c+dy} = \frac{(a+by)(c-dy)y}{(c+dy)(c-dy)y} = \frac{A}{y} + B,$$

where A and B are rational functions of x. Therefore, the calculation of the integral $\int R(x,y)dx$ reduces to the calculation of the integrals $\int B(x)dx$ and $\int \frac{A(x)dx}{y}$. The rational function $A(x)$ can be represented in the form

$$A(x) = \sum a_nx^n + \sum \frac{a_{r,m}}{(x-c_r)^m}.$$

Hence, it remains to consider the integrals of the type

$$J_n = \int \frac{x^ndx}{y} (n \geq 0) \quad \text{and} \quad H_m = \int \frac{dx}{(x-c)^my} \ (m \geq 1).$$

Since

$$\begin{aligned} \frac{d}{dx}(x^my) &= mx^{m-1}y + x^m\frac{dy}{dx} = \frac{1}{y}\left[mx^{m-1}y^2 + \frac{1}{2}x^m\frac{d(y^2)}{dx}\right] \\ &= (m+2)a_0\frac{x^{m+3}}{y} + 2(2m+3)a_1\frac{x^{m+2}}{y} \\ &\quad + 6(m+1)a_2\frac{x^{m+1}}{y} + 2(2m+1)a_3\frac{x^m}{y} + ma_4\frac{x^{m-1}}{y}, \end{aligned}$$

it follows that

$$\begin{aligned} x^my = (m+2)a_0J_{m+3} &+ 2(2m+3)a_1J_{m+2} \\ &+ 6(m+1)a_2J_{m+1} + 2(2m+1)a_3J_m + ma_4J_{m-1}. \end{aligned}$$

Using these identities for $m = 0, 1, 2, \ldots$, we can consecutively express J_3 in terms of J_0, J_1, J_2 (and a rational function of x and y), then J_4 in terms of J_0, J_1, J_2, etc. (In the case when $a_0 = 0$ we will express J_2 in terms of J_0 and J_1; next, J_3 in terms of J_0 and J_1, etc.)

To compute the integrals $H_m = \int \frac{dx}{(x-c)^m y}$, let us write the polynomial $G(x)$ in the form

$$G(x) = b_0(x-c)^4 + 4b_1(x-c)^3 + 6b_2(x-c)^2 + 4b_3(x-c) + b_4,$$

where $b_0 = a_0$. As in the preceding case, we get the identity

$$\frac{d}{dx}[(x-c)^m y] = (m+2)b_0\frac{(x-c)^{m+3}}{y} + 2(2m+3)b_1\frac{(x-c)^{m+2}}{y} + 6(m+1)b_2\frac{(x-c)^{m+1}}{y} + 2(2m+1)b_3\frac{(x-c)^m}{y} + mb_4\frac{(x-c)^{m-1}}{y}.$$

Integrating these identities for $m = -1, -2, -3, \dots$ we get

$$\begin{aligned}
\frac{y}{x-c} &= b_0\int\frac{(x-c)^2}{y}dx + 2b_1\int\frac{x-c}{y}dx - 2b_3H_1 - b_4H_2,\\
\frac{y}{(x-c)^2} &= -2b_1J_0 - 6b_2H_1 - 6b_3H_2 - 2b_4H_3,\\
\frac{y}{(x-c)^3} &= -b_0J_0 - 6b_1J_1 - 12b_2H_2 - 10b_3H_3 - 3b_4H_4,\\
\dots &\quad \dots \quad \dots \quad \dots \quad \dots \quad \dots
\end{aligned}$$

These identities enable us to express H_2, H_3, H_4, ... in terms of J_0, J_1, J_2, H_1 and rational functions of x and y. □

As we have already mentioned, any elliptic integral can be reduced to the integral $\int R(x,y)dx$, where

$$y^2 = 4x^3 - g_2x - g_3.$$

This form of elliptic integrals is called the *Weierstrass form*. Since in this case $a_0 = 0$, it follows that J_2 can be expressed in terms of J_0 and J_1; therefore, there are three types of integrals from which the rest may be calculated:

$$\int\frac{dx}{\sqrt{4x^3-g_2x-g_3}}, \quad \int\frac{xdx}{\sqrt{4x^3-g_2x-g_3}}, \quad \text{and} \quad \int\frac{dx}{(x-c)\sqrt{4x^3-g_2x-g_3}}.$$

Another widely used form of elliptic integrals is the *Legendre form*. For it the equation

$$y^2 = (1-x^2)(1-k^2x^2)$$

is used. One can pass from the Weierstrass form to the Legendre form as follows. Using linear change of variables $x = ax_1 + b$ we pass from $4x^3 - g_2x - g_3$ to $x_1(x_1-1)(x_1-k^2)$. Next, let us make the change of variables $\xi^2 = x_1^{-1}$, $\eta^2 = y^2x_1^{-3}$. We get $\eta^2 = (1-\xi^2)(1-k^2\xi^2)$.

For the Legendre form, all four types of integrals appear:

$$\int\frac{dx}{\sqrt{G(x)}}, \int\frac{xdx}{\sqrt{G(x)}}, \int\frac{x^2dx}{\sqrt{G(x)}}, \int\frac{dx}{(x-c)\sqrt{G(x)}},$$

where $G(x) = (1 - x^2)(1 - k^2x^2)$. But, since

$$\int \frac{xdx}{\sqrt{(1-x^2)(1-k^2x^2)}} = \frac{1}{2}\int \frac{du}{\sqrt{(1-u)(1-k^2u)}},$$

where $u = x^2$, this integral can be expressed in terms of elementary functions.

To simplify somewhat the form of the integrals in the Legendre form, make the change of variables $x = \sin\varphi$. Then

$$dx = \cos\varphi d\varphi, \qquad \sqrt{1-x^2} = \cos\varphi, \qquad \sqrt{1-k^2x^2} = \sqrt{1-k^2\sin^2\varphi}.$$

Therefore, the above mentioned (nonelementary) integrals take the form

$$\int \frac{d\varphi}{\sqrt{1-k^2\sin^2\varphi}}, \quad \int \frac{\sin^2\varphi d\varphi}{\sqrt{1-k^2\sin^2\varphi}}, \quad \int \frac{d\varphi}{(\sin\varphi - c)\sqrt{1-k^2\sin^2\varphi}}.$$

These integrals are called *elliptic integrals of the first*, *second*, and *third kind*, respectively.

Observe that instead of the integral $\displaystyle\int \frac{\sin^2\varphi d\varphi}{\sqrt{1-k^2\sin^2\varphi}}$ we may take the integral $\int \sqrt{1-k^2\sin^2\varphi}d\varphi$ because

$$k^2\int \frac{\sin^2\varphi\, d\varphi}{\sqrt{1-k^2\sin^2\varphi}} = \int \frac{d\varphi}{\sqrt{1-k^2\sin^2\varphi}} - \int \sqrt{1-k^2\sin^2\varphi}\, d\varphi.$$

REMARK. We abuse the language a bit when we refer to $\displaystyle\int \frac{\sin^2\varphi d\varphi}{\sqrt{1-k^2\sin^2\varphi}}$ as the elliptic integral of the second kind because in the literature this term is applied to the integral $\int \sqrt{1-k^2\sin^2\varphi}d\varphi$.

For the elliptic integrals of the first and second kind Legendre's notations are used:

$$F(\varphi) = \int_0^{\varphi} \frac{d\varphi}{\sqrt{1-k^2\sin^2\varphi}} \quad \text{and} \quad E(\varphi) = \int_0^{\varphi} \sqrt{1-k^2\sin^2\varphi}d\varphi.$$

PROBLEMS

2.6.1. Prove that a change of variables reduces the integral $\int (1+x^6)^{-1/3}dx$ to an elliptic integral.

HINT. Set $x^{-3} + x^3 = 2t^{-3/2}$.

2.6.2. Prove that a change of variables reduces the integral $\int (1-x^3)^{-2/3}dx$ to an elliptic integral.

HINT. Set $t(1-x) = (1-x^3)^{1/3}$.

§2.7. Addition theorems for the elliptic integrals $F(\varphi)$ and $E(\varphi)$

Set

$$F(\varphi) = \int_0^\varphi \frac{d\varphi}{\Delta(\varphi)} \quad \text{and} \quad E(\varphi) = \int_0^\varphi \Delta(\varphi) d\varphi,$$

where $\Delta(\varphi) = \sqrt{1 - k^2 \sin^2 \varphi}$. If $F(\varphi) + F(\psi) = F(\mu)$, then $\sin \mu$ can be algebraically expressed in terms of $\sin \varphi$ and $\sin \psi$. To prove this, consider the differential equation

$$(*) \qquad \frac{d\varphi}{\sqrt{1 - k^2 \sin^2 \varphi}} + \frac{d\psi}{\sqrt{1 - k^2 \sin^2 \psi}} = 0.$$

Its integral is $F(\varphi) + F(\psi) - F(\mu) = 0$, where μ is a constant. Taking into account that F is an odd function, the integral can be expressed in the form $F(\varphi) + F(\psi) + F(-\mu) = 0$. Let us show that the integral of the differential equation $(*)$ satisfies the relation

$$(7.1) \qquad \cos \varphi \cos \psi - \sin \varphi \sin \psi \sqrt{1 - k^2 \sin^2 \mu} = \cos \mu$$

which can be rewritten after squaring in a more symmetric form:

$$(7.2) \quad \cos^2 \varphi + \cos^2 \psi + \cos^2 \mu - 2 \cos \varphi \cos \psi \cos \mu + k^2 \sin^2 \varphi \sin^2 \psi \sin^2 \mu = 1.$$

The term "symmetric" means that not only is (7.1) satisfied but also the relations

$$(7.3) \qquad \cos \mu \cos \varphi + \sin \mu \sin \varphi \sqrt{1 - k^2 \sin^2 \psi} = \cos \psi,$$

$$(7.4) \qquad \cos \mu \cos \psi + \sin \mu \sin \psi \sqrt{1 - k^2 \sin^2 \varphi} = \cos \varphi,$$

since the arguments φ, ψ and $-\mu$ enter (7.2) symmetrically.

Divide both sides of (7.1) by $\sin \varphi \sin \psi$ and differentiate the obtained expression. The result can be expressed in the form

$$d\varphi \left(\frac{\cos \psi - \cos \mu \cos \varphi}{\sin \varphi} \right) + d\psi \left(\frac{\cos \varphi - \cos \mu \cos \psi}{\sin \psi} \right) = 0.$$

Making use of formulas (7.3) and (7.4) we get

$$\frac{d\varphi}{\sqrt{1 - k^2 \sin^2 \varphi}} + \frac{d\psi}{\sqrt{1 - k^2 \sin^2 \psi}} = 0.$$

Thus, (7.1) is indeed an integral of the differential equation $(*)$. But it cannot have two independent integrals and, therefore, the equality $F(\varphi) + F(\psi) = F(\mu)$ implies that

$$\cos \varphi \cos \psi - \sin \varphi \sin \psi \sqrt{1 - k^2 \sin^2 \mu} = \cos \mu.$$

This provides us with an implicit expression for $\cos \mu$. It is not difficult to get an explicit expression also. Let $x = \cos \mu$; then $\sin^2 \mu = 1 - x^2$. Relation (7.2) can be considered as a quadratic expression in x. By solving it for x we get

$$(7.5) \qquad \cos \mu = \frac{\cos \varphi \cos \psi - \sin \varphi \sin \psi \Delta(\varphi) \Delta(\psi)}{1 - k^2 \sin^2 \varphi \sin^2 \psi}.$$

(The sign is chosen so that the formulas (7.5) and (7.1) are compatible for small φ and ψ.)

By direct algebraic transformations we can derive from (7.5) the following expressions for $\sin\mu$ and $\Delta(\mu)$:

$$\sin\mu = \frac{\sin\varphi\cos\psi\Delta(\psi) + \sin\psi\cos\varphi\Delta(\varphi)}{1 - k^2\sin^2\varphi\sin^2\psi}, \tag{7.6}$$

$$\Delta(\mu) = \frac{\Delta(\varphi)\Delta(\psi) - k^2\sin\varphi\sin\psi\cos\varphi\cos\psi}{1 - k^2\sin^2\varphi\sin^2\psi}. \tag{7.7}$$

Dividing (7.6) by (7.5) we get

$$\tan\mu = \frac{\tan\varphi\Delta(\varphi) + \tan\psi\Delta(\psi)}{1 - \tan\varphi\tan\psi\Delta(\varphi)\Delta(\psi)}. \tag{7.8}$$

The latter formula can be interpreted as follows. Let the angles φ' and ψ' be such that $\tan\varphi' = \tan\varphi\Delta(\varphi)$ and $\tan\psi' = \tan\psi\Delta(\psi)$. Then $\mu = \varphi' + \psi'$.

In applications the case when $\mu = \frac{1}{2}\pi$ is quite important. In this case $\cos\mu = 0$ and $\sin\mu = 1$. From (7.3) and (7.4) we get $\sin\varphi = \cos\psi/\Delta(\psi)$ and $\sin\psi = \cos\varphi/\Delta(\varphi)$, and from (7.1) we get $\cos\varphi\cos\psi = b\sin\varphi\sin\psi$, i.e., $b\tan\varphi\tan\psi = 1$, where $b = \sqrt{1-k^2}$. It follows from (7.5) that

$$\cos\varphi\cos\psi = \Delta(\varphi)\Delta(\psi)\sin\varphi\sin\psi$$

and, therefore, $\Delta(\varphi)\Delta(\psi) = b$. Hence,

$$\cos\varphi = \sin\psi\Delta(\varphi) = \frac{b\sin\psi}{\Delta(\psi)} \quad \text{and} \quad \cos\psi = \frac{b\sin\varphi}{\Delta(\varphi)}.$$

Formulas (7.5) and (7.6) resemble, to some extent, the formulas for the cosine and sine of the sum of two angles. With their help we can obtain expressions similar to expressions of $\cos n\varphi$ and $\sin n\varphi$ in terms of $\cos\varphi$ and $\sin\varphi$. Let $F(\varphi_n) = nF(\varphi)$. Then

$$\begin{aligned} \sin\varphi_2 &= \frac{2\sin\varphi\cos\varphi\Delta(\varphi)}{1-k^2\sin^4\varphi}, & \cos\varphi_2 &= \frac{1-2\sin^2\varphi + k^2\sin^4\varphi}{1-k^2\sin^4\varphi}, \\ \Delta(\varphi_2) &= \frac{1-2k^2\sin^2\varphi + k^2\sin^4\varphi}{1-k^2\sin^4\varphi}, & \tan\varphi_2 &= \frac{2\tan\varphi\Delta(\varphi)}{1-(\tan\varphi\Delta(\varphi))^2}. \end{aligned}$$

To find φ from a given φ_2, we can use the fact that $\tan(\frac{\varphi_2}{2}) = \tan\varphi\Delta(\varphi)$ and may also solve the equation

$$\cos\varphi_2 = \frac{1-2x^2+k^2x^4}{1-k^2x^4},$$

where $x = \sin\varphi$. This equation corresponds to the division of $F(\varphi_2)$ in halves.

To divide $F(\psi)$ into three equal parts, we must solve the equation

$$\sin\psi = \frac{3x - 4(1+k^2)x^3 + 6k^2x^5 - k^4x^9}{1-6k^2x^4+4k^2(1+k^2)x^6 - 3k^4x^8},$$

where $x = \sin\varphi$. For $\psi = \frac{\pi}{2}$ we get the equation

$$(1+x)(1-2x+2k^2x^3-k^2x^4)^2 = 0,$$

i.e., the division of $F(\pi/2)$ into three equal parts reduces to the solution of the equation

$$1 - 2\sin\varphi + 2k^2\sin^3\varphi + k^2\sin^4\varphi = 0.$$

For the elliptic integral of the second kind, $E(\varphi)$, there is only an addition theorem with an extra algebraic term. It is directly connected with the addition theorem for the elliptic integral of the first kind, $F(\varphi)$.

THEOREM. *If* $F(\varphi)+F(\psi)-F(\mu)=0$, *then*

$$E(\varphi)+E(\psi)-E(\mu)=k^2\sin\varphi\sin\psi\sin\mu.$$

PROOF. Let $E(\varphi)+E(\psi)-E(\mu)=P(\varphi,\psi,\mu)$. Let us differentiate this equality for a constant value of μ.

As a result we get

$$\Delta(\varphi)d\varphi+\Delta(\psi)d\psi=dP.$$

But by (7.3) and (7.4)

$$\Delta(\varphi)=\frac{\cos\varphi-\cos\psi\cos\mu}{\sin\psi\sin\mu}\quad\text{and}\quad\Delta(\psi)=\frac{\cos\psi-\cos\varphi\cos\mu}{\sin\varphi\sin\mu}.$$

Hence,

$$\begin{aligned}dP&=\left(\frac{\cos\varphi-\cos\psi\cos\mu}{\sin\psi\sin\mu}\right)d\varphi+\left(\frac{\cos\psi-\cos\varphi\cos\mu}{\sin\varphi\sin\mu}\right)d\psi\\&=\frac{d(\sin^2\varphi+\sin^2\psi+2\cos\varphi\cos\psi\cos\mu)}{2\sin\varphi\sin\psi\sin\mu}.\end{aligned}$$

By (7.2) we have

$$1-\sin^2\varphi+1-\sin^2\psi-2\cos\varphi\cos\psi\cos\mu=1-\cos^2\mu-k^2\sin^2\varphi\sin^2\psi\sin^2\mu.$$

Hence,

$$dP=\frac{d(k\sin\varphi\sin\psi\sin\mu)^2}{2\sin\varphi\sin\psi\sin\mu}=k^2d(\sin\varphi\sin\psi\sin\mu).$$

Since both P and $\sin\varphi\sin\psi\sin\mu$ vanish at $\varphi=0$, we get $P=k^2\sin\varphi\sin\psi\sin\mu$. □

This addition theorem enables us to get the following expressions for $nE(\varphi)-E(\varphi_n)$, where $F(\varphi_n)=nF(\varphi)$:

$$\begin{aligned}2E(\varphi)-E(\varphi_2)&=k^2\sin\varphi\sin\varphi\sin\varphi_2,\\3E(\varphi)-E(\varphi_3)&=(2E(\varphi)-E(\varphi_2))+(E(\varphi_2)+E(\varphi)-E(\varphi_3))\\&=k^2\sin\varphi(\sin\varphi\sin\varphi_2+\sin\varphi_2\sin\varphi_3),\end{aligned}$$

etc. In Chapter 3 we will apply the formulas obtained in this section to solve various problems on arc lengths of an ellipse.

§2.8. The elliptic Jacobi functions

The function considered in the preceding section

$$F(\varphi)=\int_0^x\frac{dx}{\sqrt{(1-x^2)(1-k^2x^2)}}=\int_0^\varphi\frac{d\varphi}{\sqrt{1-k^2\sin^2\varphi}},$$

where $x=\sin\varphi$, is in many ways analogous to the function

$$\arcsin x=\int_0^x\frac{dx}{\sqrt{1-x^2}}=\int_0^\varphi d\varphi,$$

where $x = \sin\varphi$. For example, if $F(\varphi) + F(\psi) = F(\mu)$, then $\sin\mu$ is expressed in terms of $\sin\varphi$ and $\sin\psi$ by the formula

$$\sin\mu = \frac{\sin\varphi\cos\psi\Delta(\psi) + \sin\psi\cos\psi\Delta(\varphi)}{1 - k^2\sin^2\varphi\sin^2\psi}, \tag{8.1}$$

and if $\arcsin x + \arcsin y = \arcsin z$, where $x = \sin\varphi$, $y = \sin\psi$, and $z = \sin\mu$, then

$$\sin\mu = \sin\varphi\cos\psi + \sin\psi\cos\varphi. \tag{8.2}$$

This analogy is not accidental: for $k = 0$ the function $F(\varphi)$ becomes the function $\arcsin x = \varphi$ and formula (8.1) turns into (8.2).

For many reasons it is more convenient to consider not the function $\varphi = \arcsin x$ but the inverse function $x = \sin\varphi$. The function inverse to the function $F(\varphi)$ is also in many respects more convenient than the function $F(\varphi)$ itself. Let us replace Legendre's notation $F(\varphi)$ with the Jacobi notation: set $u(\varphi) = F(\varphi)$. The function $\varphi(u)$ inverse to $u(\varphi)$ is called the *amplitude* of u and is denoted by $\varphi = \operatorname{am} u$. In the preceding section we obtained formulas (7.5)–(7.7) for the functions $\sin\varphi$, $\cos\varphi$ and $\Delta(\varphi) = \sqrt{1 - k^2\sin^2\varphi}$. Introducing the functions

$$\operatorname{sn} u = \sin\operatorname{am} u,\ \operatorname{cn} u = \cos\operatorname{am} u \quad\text{and}\quad \operatorname{dn} u = \Delta(\operatorname{am} u)$$

we may express formulas (7.5)–(7.7) in the following way:

$$\operatorname{cn}(u+v) = \frac{\operatorname{cn} u\operatorname{cn} v - \operatorname{sn} u\operatorname{sn} v\operatorname{dn} u\operatorname{dn} v}{1 - k^2\operatorname{sn}^2 u\operatorname{sn}^2 v}, \tag{8.3}$$

$$\operatorname{sn}(u+v) = \frac{\operatorname{sn} u\operatorname{cn} v\operatorname{dn} v + \operatorname{sn} v\operatorname{cn} u\operatorname{dn} u}{1 - k^2\operatorname{sn}^2 u\operatorname{sn}^2 v}, \tag{8.4}$$

$$\operatorname{dn}(u+v) = \frac{\operatorname{dn} u\operatorname{dn} v - k^2\operatorname{sn} u\operatorname{sn} v\operatorname{cn} u\operatorname{cn} v}{1 - k^2\operatorname{sn}^2 u\operatorname{sn}^2 v}. \tag{8.5}$$

The functions $\operatorname{sn} u$, $\operatorname{cn} u$ and $\operatorname{dn} u$ are usually called *elliptic Jacobi functions* although many properties of these functions had been established, to some extent, by Legendre and Abel before Jacobi.

One of the most important properties of the functions $\operatorname{sn} u$, $\operatorname{cn} u$ and $\operatorname{dn} u$ is their double periodicity. The existence of one period for these functions is quite obvious. Indeed, the functions $\sin\varphi$ and $\cos\varphi$ have period $2\pi = 4(\frac{\pi}{2})$ and the function $\sin^2\varphi$ has period $\pi = 2(\frac{\pi}{2})$. Therefore, the functions $\operatorname{sn} u$ and $\operatorname{cn} u$ have period $4K$, where

$$K = \int_0^{\pi/2} \frac{d\varphi}{\sqrt{1 - k^2\sin^2\varphi}};$$

and the function $\operatorname{dn} u = \sqrt{1 - k^2\operatorname{sn}^2 u}$ has period $2K$.

With the help of the addition formulas (8.3)–(8.5) we can figure out how the functions $\operatorname{sn} u, \operatorname{cn} u$ and $\operatorname{dn} u$ behave if the argument is increased by a quarter of the period, K, and by a half period, $2K$. Substituting

$$\operatorname{sn} K = 1,\quad \operatorname{cn} K = 0 \quad\text{and}\quad \operatorname{dn} K = \sqrt{1 - k^2}$$

into (8.3)–(8.5) we get

$$\operatorname{sn}(u+K) = \frac{\operatorname{cn} u}{\operatorname{dn} u},\quad \operatorname{cn}(u+K) = -\frac{\sqrt{1-k^2}\operatorname{sn} u}{\operatorname{dn} u},\quad \operatorname{dn}(u+K) = \frac{\sqrt{1-k^2}}{\operatorname{dn} u}.$$

Since $\operatorname{sn} 2K = 0$, $\operatorname{cn} 2K = -1$ and $\operatorname{dn} 2K = 1$, it follows that

$$\operatorname{sn}(u+2K) = -\operatorname{sn} u, \ \operatorname{cn}(u+2K) = -\operatorname{cn} u, \ \operatorname{dn}(u+2K) = \operatorname{dn} u.$$

It is somewhat more difficult to figure out what the other period of the elliptic Jacobi functions is. Let us first recall that the integral

$$\int \frac{d\varphi}{\sqrt{1-k^2\sin^2\varphi}}$$

was obtained from the integral

$$\int \frac{dx}{\sqrt{(1-x^2)(1-k^2x^2)}}$$

with the help of the change of variables $x = \sin\varphi$. It will be more convenient now to work with the original integral. On the complex plane $\mathbb{C}$, the function

$$u(x) = \int_0^x \frac{dx}{\sqrt{(1-x^2)(1-k^2x^2)}}$$

is not defined in general because the value of $u(x)$ depends on the contour of integration. The values of the function $u(x)$ at the same point can differ by numbers of the form

$$L = \int_C \frac{dx}{\sqrt{(1-x^2)(1-k^2x^2)}},$$

where the integral is taken along a closed contour C. Here any such number L is a period of the inverse function $x(u)$. The integrand has singular points ± 1, $\pm k^{-1}$.

Let us see how the choice of the path around these points affects the values of the functions $\operatorname{sn} u = x, \operatorname{cn} u = \sqrt{1-x^2}$ and $\operatorname{dn} u = \sqrt{1-k^2x^2}$.

Let

$$\int_0^1 \frac{dx}{\sqrt{(1-x^2)(1-k^2x^2)}} = K \quad \text{and} \quad \int_0^X \frac{dx}{\sqrt{(1-x^2)(1-k^2x^2)}} = \alpha.$$

The path shown in Figure 22 shows that the values of the integral at the point X are equal to α and $K + (K-\alpha) = 2K - \alpha$. Indeed, on the last part of the path both the sign of the function $\sqrt{1-x^2}$ and the direction of the segment of integration change; as the result of two such changes the sign of the integral does not change. Therefore, $\operatorname{sn}\alpha = \operatorname{sn}(2K-\alpha)$. Furthermore, the sign of the function $\sqrt{1-x^2}$ changes under this passage and the sign of $\sqrt{1-k^2x^2}$ does not change. Hence, $\operatorname{cn}\alpha = -\operatorname{cn}(2K-\alpha)$ and $\operatorname{dn}\alpha = \operatorname{dn}(2K-\alpha)$. Replacing α with $-\alpha$ we get

$$\begin{aligned}
\operatorname{sn}(\alpha+2K) &= \operatorname{sn}(-\alpha) = -\operatorname{sn}\alpha,\\
\operatorname{cn}(\alpha+2K) &= -\operatorname{cn}(-\alpha) = -\operatorname{cn}\alpha,\\
\operatorname{dn}(\alpha+2K) &= \operatorname{dn}(-\alpha) = \operatorname{dn}\alpha.
\end{aligned}$$

We have already obtained these formulas by another method.

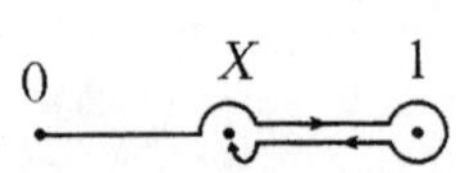

FIGURE 22

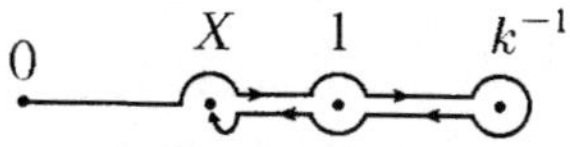

FIGURE 23

Now, let us consider the passage along the curve shown in Figure 23. Let

$$\int_1^{k^{-1}} \frac{dx}{\sqrt{1-x^2}\sqrt{1-k^2x^2}} = iK'.$$

(The number K' is real since the number $\sqrt{1-x^2}$ is purely imaginary for $x \in (1, k^{-1})$.) Therefore, the values of the integral at point X are equal to α and $K+iK'+iK'-(K-\alpha)$. Here only the sign of the last summand needs elucidation.

Observe that there occurred three changes of sign: the direction of the integration contour had changed, and so did the signs of both functions $\sqrt{1-x^2}$ and $\sqrt{1-k^2x^2}$. As a result, we get

$$\begin{aligned}\operatorname{sn}\alpha &= \operatorname{sn}(\alpha+2iK'),\\ \operatorname{cn}\alpha &= -\operatorname{cn}(\alpha+2iK'),\\ \operatorname{dn}\alpha &= -\operatorname{dn}(\alpha+2iK').\end{aligned}$$

Therefore, the functions $\operatorname{sn} u$, $\operatorname{cn} u$, and $\operatorname{dn} u$ have periods $2iK'$, $4iK'$, and $4iK'$, respectively. Moreover, the function $\operatorname{cn} u$ has a period $2K+2iK'$. Indeed,

$$\operatorname{cn}(\alpha+2iK'+2K) = -\operatorname{cn}(\alpha+2iK') = \operatorname{cn}\alpha.$$

Thus, the function $\operatorname{sn} u$ has periods $4K$ and $2iK$; the function $\operatorname{cn} u$ has periods $4K$ and $2K+2iK'$; and the function $\operatorname{dn} u$ has periods $2K$ and $4iK'$.

PROBLEMS

2.8.1. Prove that $K' = \int_0^1 \frac{dt}{\sqrt{(1-t^2)(1-k'^2t^2)}}$, where $k'^2+k^2=1$.

HINT. Make the change $k't = \sqrt{1-k^2x^2}$, where k' is chosen so that $k\sqrt{x^2-1} = k'\sqrt{1-t^2}$.

§2.9. The Weierstrass theorem on functions possessing an algebraic addition theorem

We will say that a meromorphic function $\varphi(z)$ *possesses an algebraic addition theorem* if there exists a nonzero polynomial F in three variables such that

$$F(\varphi(z_1+z_2), \varphi(z_1), \varphi(z_2)) = 0;$$

this identity means that $\varphi(z_1+z_2)$ can be algebraically expressed in terms of $\varphi(z_1)$ and $\varphi(z_2)$.

For example, the function $\operatorname{sn} z$ possesses an algebraic addition theorem. Indeed, if $a = \operatorname{sn}(z_1+z_2), b = \operatorname{sn} z_1$ and $c = \operatorname{sn} z_2$, then

$$a = \frac{b\sqrt{1-c^2}\sqrt{1-k^2c^2} + c\sqrt{1-b^2}\sqrt{1-k^2b^2}}{1-k^2b^2c^2}.$$

Squaring twice we can get rid of the radicals and get a polynomial relation

$$F(a,b,c)=0.$$

The Weierstrass function $\wp(z)$ also possesses an algebraic addition theorem. Indeed, if $a=\wp(z_1+z_2)$, $b=\wp(z_1)$ and $c=\wp(z_2)$, then

$$a=-b-c+\frac{1}{4}\left(\frac{\sqrt{4b^3-g_2b-g_3}-\sqrt{4c^3-g_2c-g_3}}{b-c}\right)^2.$$

By simplifying this formula we get a relation of the form $F(a,b,c)=0$, where F is a polynomial.

As we saw in §2.3, any elliptic function f can be represented in the form

$$f=R(\wp)+R_1(\wp)\wp',$$

where R and R_1 are rational functions (see §2.3). This representation enables us to get an algebraic addition theorem for an arbitrary elliptic function if we take into account that

$$(\wp')^2=4\wp^3-g_2\wp-g_3.$$

Indeed, if $R_1\neq 0$, then $\wp'=\frac{f-R(\wp)}{R_1(\wp)}$. Hence,

$$\left(\frac{f-R(\wp)}{R_1(\wp)}\right)^2=4\wp^3-g_2\wp-g_3.$$

This relation can be rewritten in the form $P(f,\wp)=0$, where P is a polynomial (its degree with respect to f is equal to 2). If $R_1=0$, the relation $f=R(\wp)$ can also be expressed in the above form. Let $A=f(z_1+z_2)$, $B=f(z_1)$, and $C=f(z_2)$. From the relations $F(a,b,c)=0$ and $P(A,a)=0$ we can get a relation $F_1(A,b,c)=0$ by calculating the resultant of the polynomials $f(a)=F(a,b,c)$ and $g(a)=P(A,a)$.

Next, from the relations $F_1(A,b,c)=0$ and $P(B,b)=0$ we get $F_2(A,B,c)=0$ and from the relations $F_2(A,B,c)=0$ and $P(C,c)=0$ we get a relation $G(A,B,C)=0$, as required.

Not only elliptic functions possess algebraic addition theorems. For instance, the *exponential* function e^z possesses the algebraic addition theorem, since $e^{z_1+z_2}=e^{z_1}e^{z_2}$. Moreover, if $u=R(f)$, where R is a rational function, then $P(u,f)=0$, where P is a polynomial (linear in u). Hence, from the relation $F(a,b,c)=0$, where

$$F(a,b,c)=\begin{cases} a-(b+c) & \text{if } f=z,\\ a-bc & \text{if } f=e^{\lambda z},\end{cases}$$

we may get, as above, a relation $G(A,B,C)=0$, where $A=R(a)$, $B=R(b)$, and $C=R(c)$. Therefore, any rational function and also any rational function in $e^{\lambda z}$ possesses an algebraic addition theorem.

It turns out that the above examples exhaust all meromorphic functions possessing an algebraic addition theorem.

THEOREM (Weierstrass). *Any meromorphic function $\varphi(z)$ possessing an algebraic addition theorem is either an elliptic function or is of the form $R(z)$ or $R(e^{\lambda z})$, where R is a rational function.*

PROOF (W. S. Osgood). In the finite domain the meromorphic function has no singular points other than poles. If the limits

$$\lim_{z\to\infty}\varphi(z) \quad \text{or} \quad \lim_{z\to\infty}\frac{1}{\varphi(z)}$$

exist, then the function $\varphi(z)$ is rational. Indeed, let us subtract from φ the sum of its principal parts at all the poles (if the point ∞ is a pole, then the principal part of the function at this point is of the form $a_r z^r + a_{r+1}z^{z+1} + \cdots$, where $r > 0$). As a result, we get a function f without singular points; the point ∞ is also a nonsingular one. Therefore, f is a constant and the initial function φ is rational.

In what follows we will assume that the function φ is not rational, i.e., the point ∞ is an essential singularity. To prove the Weierstrass theorem, we will need the following theorem on the behavior of a function in a neighborhood of an essentially singular point.

PICARD'S BIG THEOREM. *Any analytic function $\varphi(z)$ assumes in an arbitrary neighborhood of an essentially singular point any finite value except, perhaps, one value.*

Proof of this theorem can be found, e.g., in [**B16**].

Let $F(\varphi(z_1+z_2), \varphi(z_1), \varphi(z_2)) = 0$, where F is a polynomial whose degree with respect to the first argument is equal to n. We have to prove that φ is a periodic function and if it is not doubly periodic, then $\varphi(z) = R(e^{\lambda z})$. Picard's big theorem implies that in a neighborhood of an essentially singular point the function φ takes a certain value c infinitely many times.

Let $a_1, \dots, a_{n+1}$ be points at which φ takes value c. The singular points of φ, together with the points z for which the points $z + a_i$ are singular, form a set of measure zero. Therefore, there exists a nonsingular point z_0 of the function φ for which all points $a_i + z_0$ are also nonsingular ones. Then the points z and $a_i + z$ for values z sufficiently close to z_0 are also nonsingular. For such points z consider the equation

$$F(x, \varphi(z), c) = 0. \tag{9.1}$$

It has $n + 1$ roots $x_i = \varphi(z + a_i)$, because

$$F(\varphi(z + a_i), \varphi(z), c) = F(\varphi(z + a_i), \varphi(z), \varphi(a_i)) = 0.$$

Equation (9.1) is a nonzero polynomial in x of degree n and, therefore, it has at most n distinct roots. It follows that $\varphi(z + a_p) = \varphi(z + a_q)$ for certain distinct p and q. Such a relation is satisfied for any point z from a neighborhood of z_0, but the pairs (p, q) can differ.

Nevertheless, there are only finitely many pairs (p, q) and, therefore, some relation $\varphi(z + a_p) = \varphi(z + a_q)$ holds for an infinite set of points z. These points have a limit point z_1 and the function φ is regular at z_1. By the uniqueness theorem we see that the functions $\varphi(z + a_p)$ and $\varphi(z + a_q)$ coincide, i.e., $a_p - a_q$ is a period of φ.

We have proved that φ is a periodic function; for definiteness, we may assume that the minimal period of φ is equal to 2π. Let us assume that φ has no other periods and then show that $\varphi(z) = R(w)$, where $w = e^{iz}$ and R is a rational function. The map $z \longmapsto w = e^{iz}$ sends the strip $0 \le \operatorname{Re} z < 2\pi$ into the plane with the cut from 0 to $+\infty$. The function $\psi(w) = \varphi(z)$ is meromorphic on the plane with points 0 and ∞ punctured. If these points are not essentially singular ones, then the function ψ is rational.

Suppose that at least one of these poles is essentially singular. Then by Picard's big theorem there exist points $b_1, \dots, b_{n+1}$ for which $\psi(b_i) = c$. The preimages $\beta_1, \dots, \beta_{n+1}$ of these points with respect to the map $z \longmapsto w$ are distinct and belong to the strip $0 \le \operatorname{Re} z < 2\pi$. The equation $F(x, \varphi(z), c) = 0$ has roots $x_i = \varphi(\beta_i + z)$. Repeating the same arguments as above, we see that the function φ has a period $\beta_p - \beta_q$, where $0 \le \operatorname{Re}\beta_p, \operatorname{Re}\beta_q < 2\pi$. Therefore, either φ has another period in addition to the purely real period 2π or it has a real period smaller than 2π. The latter case is impossible, since the period 2π is the smallest one by the assumption. □

CHAPTER 3

Arcs of Curves and Elliptic Integrals

For a circle it is easy to construct an arc whose length is equal to the sum of lengths of two other arcs of this circle. Generally speaking, this is related to the fact that $\sin(\varphi + \psi)$ is expressed in terms of $\sin\varphi$ and $\sin\psi$ by the formula

$$\sin(\varphi + \psi) = \sin\varphi\sqrt{1 - \sin^2\psi} + \sin\psi\sqrt{1 - \sin^2\varphi}.$$

For the elliptic integrals $F(\varphi)$ and $E(\varphi)$ there also exist addition theorems, although more cumbersome, as we have seen. This means that for curves whose arc lengths are expressed in terms of $F(\varphi)$ and $E(\varphi)$ the operation of addition of arc lengths is also possible, although it is more involved; moreover, for $E(\varphi)$ it is not, strictly speaking, the straightforward addition, but an addition with a certain extra algebraic term.

In this chapter we will study curves whose arc lengths can be expressed in terms of elliptic integrals. In many respects the lemniscate is the most interesting of all such curves. However, the lemniscate deserves special study and we will study it in the next chapter.

§3.1. Arcs of the ellipse and the hyperbola

The ellipse $\frac{x^2}{a^2} + \frac{y^2}{b^2} = 1$ can be given parametrically by the formulas $x = a\cos\varphi$, $y = b\sin\varphi$. The differential dl of the length of an arc on the ellipse is equal to $\sqrt{dx^2 + dy^2} = d\varphi\sqrt{a^2\cos^2\varphi + b^2\sin^2\varphi}$. If $a = 1$ and $b = \sqrt{1 - k^2}$, then $dl = d\varphi\sqrt{1 - k^2\sin^2\varphi}$. In this case the length of the arc on the ellipse between the end point of the small half axis, B, and the point $M = (\cos\varphi, b\sin\varphi)$ is equal to (see Figure 24)

$$E(\varphi) = \int_0^{\varphi} \sqrt{1 - k^2\sin^2\psi}\, d\psi.$$

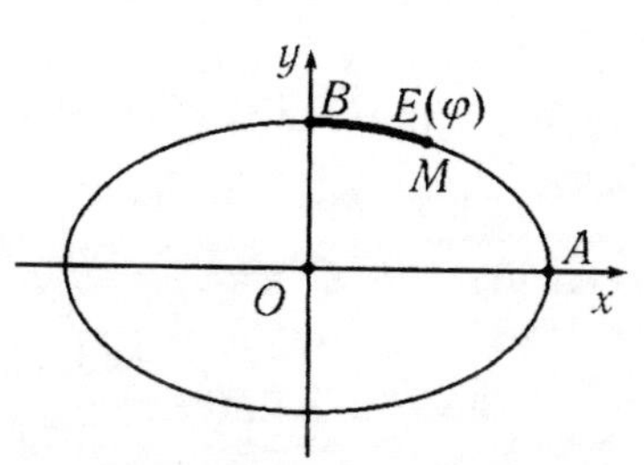

FIGURE 24

Therefore, the length of an arc on the ellipse can be expressed in terms of an elliptic integral of the second kind. This is precisely the reason why the integral gets the name *elliptic*.

The simplest parameterization of the hyperbola $\frac{x^2}{a^2} - \frac{y^2}{b^2} = 1$ is obtained using hyperbolic functions $x = a\cosh t$, $y = b\sinh t$. To express the length of an arc on the hyperbola in terms of $F(\varphi)$ and $E(\varphi)$ we, however, need a parameterization of the hyperbola by trigonometric functions. One such parameterization is given by the formulas

$$x = \frac{a}{\cos\varphi}, \quad y = b\tan\varphi.$$

Under such a parameterization the differential of the arc length is equal to

$$\frac{1}{\cos^2\varphi}\sqrt{a^2\sin^2\varphi + b^2}\, d\varphi$$

and this formula does not lead to the expression desired. Therefore, let us consider another parameterization setting $y = b^2\tan\varphi$. Then

$$x^2 = \left(\frac{a}{\cos\varphi}\right)^2 (1 - (1 - b^2)\sin^2\varphi).$$

In particular, if $a^2 = 1 - b^2 = k^2$ we get

$$x = \frac{k}{\cos\varphi}\sqrt{1 - k^2\sin^2\varphi} \quad \text{and} \quad y = (1 - k^2)\tan\varphi$$

so that

$$dl = \frac{(1 - k^2)d\varphi}{\cos^2\varphi\sqrt{1 - k^2\sin^2\varphi}}.$$

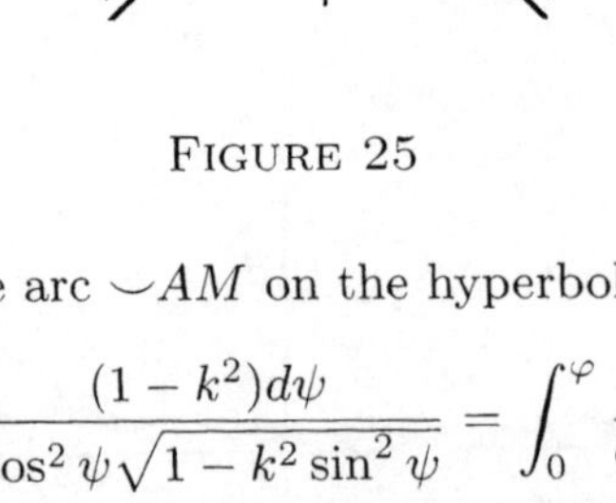

FIGURE 25

Hence, the length of the arc $\smile AM$ on the hyperbola (Figure 25) is equal to

$$\Gamma(\varphi) = \int_0^\varphi \frac{(1 - k^2)d\psi}{\cos^2\psi\sqrt{1 - k^2\sin^2\psi}} = \int_0^\varphi \frac{(1 - k^2)d\psi}{\cos^2\psi\Delta(\psi)},$$

where $\Delta(\psi) = \sqrt{1 - k^2\sin^2\psi}$. Since $\Delta'(\psi) = -\dfrac{k^2\sin\psi\cos\psi}{\Delta(\psi)}$, it follows that

$$(\Delta(\psi)\tan\psi)' = -\frac{k^2\sin^2\psi}{\Delta(\psi)} + \frac{\Delta(\psi)}{\cos^2\psi} = \frac{1 - k^2}{\cos^2\psi\Delta(\psi)} - \frac{1 - k^2}{\Delta(\psi)} + \Delta(\psi)$$

so that

$$\begin{aligned}\Gamma(\varphi) &= \Delta(\varphi)\tan\varphi - \int_0^{\varphi} \Delta(\psi)d\psi + (1-k^2)\int_0^{\varphi} \frac{d\psi}{\Delta(\psi)} \\ &= \Delta(\varphi)\tan\varphi - E(\varphi) + (1-k^2)F(\varphi).\end{aligned}$$

Thus, the arc length of the hyperbola can also be expressed in terms of the elliptic integrals $E(\varphi)$ and $F(\varphi)$ (and the elementary function $\Delta(\varphi)\tan\varphi$), although this expression is more cumbersome.

§3.2. Division of arcs of the ellipse

Let us consider the ellipse

$$\frac{x^2}{a^2} + \frac{y^2}{b^2} = 1,$$

where $a = 1$ and $b = \sqrt{1-k^2}$. The length of its arc $\smile BM$ (Figure 26) is equal to $E(\varphi)$. In §2.7 we showed that if $F(\varphi) + F(\psi) = F(\mu)$, then

$$E(\varphi) + E(\psi) - E(\mu) = k^2 \sin\varphi \sin\psi \sin\mu,$$

where φ, ψ, and μ are subject to the constraint

$$\cos\varphi\cos\psi - \sin\varphi\sin\psi\sqrt{1-k^2\sin^2\mu} = \cos\mu.$$

We do not repeat all the formulas from §2.7, although we will need many of them now to obtain relations between the lengths of the ellipse arcs of the ellipse.

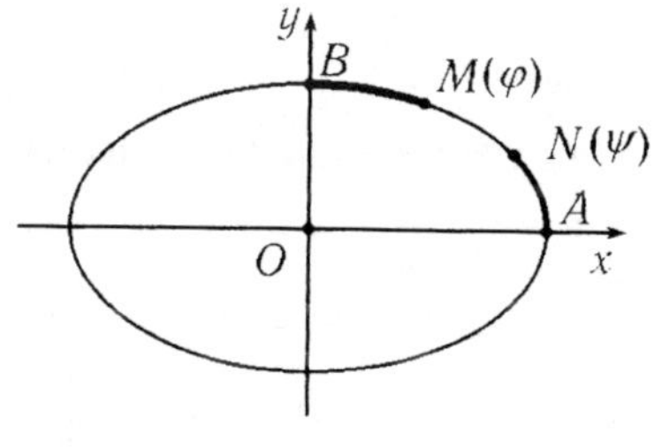

FIGURE 26

Let us start with the study of the case $\mu = \pi/2$. In this case $\sin\mu = 1$ and the angles φ and ψ are related by the formula $b\tan\varphi\tan\psi = 1$. Let the points M and N correspond to the angles φ and ψ (see Figure 26). Denote the length of the arc $\smile BM$ by $\overset{\frown}{BM}$. Then

$$E(\varphi) = \overset{\frown}{BM}, \quad E(\psi) = \overset{\frown}{BM} + \overset{\frown}{MN}, \quad E(\mu) = \overset{\frown}{BM} + \overset{\frown}{MN} + \overset{\frown}{NA}.$$

Therefore,

$$E(\varphi) + E(\psi) - E(\mu) = \overset{\frown}{BM} - \overset{\frown}{NA}.$$

As a result, we get the following statement.

THEOREM (Fagnano). *Let the angles φ and ψ be related by the formula*

$$b \tan\varphi \tan\psi = 1,$$

and let M and N be the points of the ellipse corresponding to these angles. Then

$$\overset{\frown}{BM} - \overset{\frown}{NA} = k^2 \sin\varphi \sin\psi.$$

(Recall that $a = 1$ and $b = \sqrt{1-k^2}$.)

Let us consider in detail the case when the points M and N coincide, i.e., $\varphi = \psi = \theta$. Then $\tan^2\theta = b^{-1}$ and $\sin^2\theta = (1+b)^{-1}$. Hence, $\overset{\frown}{BM} - \overset{\frown}{AM} = k^2\sin^2\theta = 1-b$ and $\overset{\frown}{BM} + \overset{\frown}{AM} = E^1 = E(\frac{\pi}{2})$ is the quarter of the arc length of the ellipse. Therefore, $\overset{\frown}{BM} = \frac{1}{2}E^1 + \frac{1}{2}(1-b)$ and $\overset{\frown}{AM} = \frac{1}{2}E^1 - \frac{1}{2}(1-b)$, i.e., up to an extra algebraic term $\frac{1}{2}(1-b)$ the point M divides the arc by half. The presence of the extra term $k^2 \sin\varphi \sin\psi \sin\mu$ in the addition theorem for elliptic integrals of the second kind leads to this, not quite natural, division "by half".

A similar division of the arc "by half" can also be performed not only for the quarter of the arc of the ellipse but for an arbitrary arc $\smile BM$. If $\varphi = \psi = \theta$ and $F(\varphi) + F(\psi) = F(\mu)$, then

$$2E(\theta) - E(\mu) = k^2\sin^2\theta \sin\mu,$$

where the angles θ and μ are related by the formula

$$\cos^2\theta - \sin^2\theta\Delta(\mu) = \cos\mu.$$

To find the angle μ from a given angle θ (this problem is called the *duplication of the arc*) we can make use of any of the formulas

$$\sin\mu = \frac{2\sin\theta\cos\theta\Delta(\theta)}{1-k^2\sin^4\theta} \quad \text{or} \quad \tan\left(\frac{\mu}{2}\right) = \Delta(\theta)\tan\theta.$$

To find the angle θ starting with a given angle μ (the division of the arc "by half") we can use the formula

$$\sin^2\theta = \frac{1-\cos\mu}{1+\Delta(\mu)}.$$

The equation

$$2E(\theta) - E(\mu) = k^2\sin^2\theta\sin\mu$$

means that

$$2\overset{\frown}{BM} - \overset{\frown}{BN} = k^2\sin^2\theta\sin\mu,$$

where the points M and N correspond to the angles θ and μ. Hence,

$$\overset{\frown}{BM} = \frac{1}{2}\overset{\frown}{BN} + \frac{1}{2}k^2\sin^2\theta\sin\mu = \frac{1}{2}\overset{\frown}{BN} + \frac{1-\Delta(\mu)}{2}\tan\frac{\mu}{2}.$$

Suppose we want to construct an arc $\smile MN$ whose length should be equal to a half of $\overset{\frown}{AB}$ (Figure 27). Then the problem of the division of the arc can be solved explicitly without an extra algebraic term.

First, construct the point K that divides $\smile AB$ "by half" and is expressed with an extra algebraic term (see above). For the angle θ corresponding to K we have

$$\sin^2\theta = (1+b)^{-1} \text{ and } E(\theta) = \frac{1}{2}E^1 + \frac{1}{2}(1-b).$$

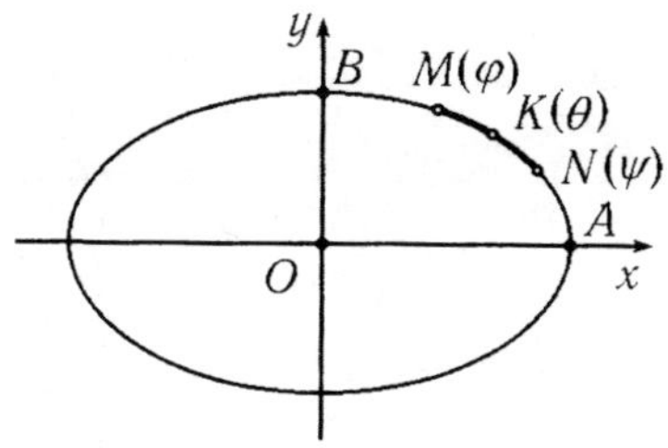

FIGURE 27

For the angles φ and ψ the relation

$$E(\psi) - E(\varphi) = \frac{1}{2}E^1$$

should be satisfied. Using the freedom in the choice M, let us impose on the angles φ and ψ an additional constraint

$$F(\varphi) + F(\theta) - F(\psi) = 0$$

from which it follows that

$$E(\varphi) + E(\theta) - E(\psi) = k^2 \sin\varphi \sin\psi \sin\theta$$

and

$$\cos\varphi\cos\psi + \sin\varphi\sin\psi\Delta(\theta) = \cos\theta.$$

Then

$$\begin{aligned}\frac{1}{2}E^1 &= E(\psi) - E(\varphi) = E(\theta) - k^2\sin\varphi\sin\psi\sin\theta \\ &= \frac{1}{2}E^1 + \frac{1}{2}(1-b) - k^2 \sin\varphi\sin\psi\sin\theta,\end{aligned}$$

so that

$$k^2\sin\varphi\sin\psi\sin\theta = \frac{1}{2}(1-b).$$

Taking into account that $k^2 = 1 - b^2$ and $\sin^2\theta = (1+b)^{-1}$, we get

$$\sin\varphi\sin\psi = \frac{(1-b)\sqrt{1+b}}{2(1-b^2)} = \frac{1}{2\sqrt{1+b}} = \frac{1}{2}\sin\theta.$$

Since $\Delta(\theta) = \sqrt{b}$, the relation

$$\cos\varphi\cos\psi + \sin\varphi\sin\psi\Delta(\theta) = \cos\theta$$

can be rewritten in the form

$$\cos\varphi\cos\psi + \frac{1}{2}\sin\theta\sqrt{b} = \cos\theta.$$

But $\sin\theta\sqrt{b} = \cos\theta$; hence, $\cos\varphi\cos\psi = \frac{1}{2}\cos\theta$. As a result, we get equations for φ and ψ:

$$\sin\varphi\sin\psi = \frac{1}{2}\sin\theta \quad \text{and} \quad \cos\varphi\cos\psi = \frac{1}{2}\cos\theta.$$

It follows that

$$\cos(\varphi \pm \psi) = \frac{1}{2}(\cos\theta \mp \sin\theta) = \cos\frac{\pi}{4}\cos\left(\theta \pm \frac{\pi}{4}\right).$$

These formulas enable us to find φ and ψ. Moreover, from the same equations we can also derive expressions

$$\sin\varphi = \frac{1}{4}\sqrt{3+4\sin\theta+2\sin^2\theta} - \frac{1}{4}\sqrt{3-4\sin\theta+2\sin^2\theta},$$
$$\sin\psi = \frac{1}{4}\sqrt{3+4\sin\theta+2\sin^2\theta} + \frac{1}{4}\sqrt{3-4\sin\theta+2\sin^2\theta}.$$

These expressions show that given an ellipse in which the major and minor semi-axes OA and OB are drawn, then with a ruler and compass we can construct an arc of the ellipse whose length is equal to a half length of the arc $\smile AB$.

Should the semi-axes not be given, they still can be constructed with a ruler and compass.

This can be done, for example, as follows. First, let us construct a pair of *conjugate* diameters of the ellipse. To this end we can use the fact that the midpoints of parallel chords of the ellipse lie on one diameter AA' of the ellipse, while the midpoints of chords parallel to AA' lie on the diameter BB' conjugate to AA'. Let O be the center of the ellipse. Let us draw the perpendicular from B on the straight line OA and on the perpendicular mark points P and Q such that $BP = BQ = OA$. Then the bisectors of the angles between the straight lines OP and OQ determine the *principal axes* of the ellipse we were looking for.

Indeed, let us consider the coordinate system whose axes are directed along the principal axes of the ellipse. Then the coordinates of the points A and B are $(a\cos\varphi, b\sin\varphi)$ and $(a\sin\varphi, -b\cos\varphi)$, respectively. Therefore, the coordinates of the points P and Q are $((a+b)\sin\varphi, -(a+b)\cos\varphi)$ and $((a-b)\sin\varphi, (a-b)\cos\varphi)$, respectively. It is clear now that the bisectors of the angles between the straight lines OP and OQ are the coordinates axes.

Now, let us investigate *the division of an arc of the ellipse into three equal parts.* First of all, consider the division with an extra algebraic term. Recall that if $F(\psi) = 3F(\varphi)$, then

$$\sin\psi = \frac{3x - 4(1+k^2)x^3 + 6k^2x^5 - k^4x^9}{1 - 6k^2x^4 + 4k^2(1+k^2)x^6 - 3k^4x^8},$$

where $x = \sin\varphi$; for $\psi = \frac{\pi}{2}$ this equation reduces to the equation

$$1 - 2x + 2k^2x^3 - k^2x^4 = 0. \tag{2.1}$$

If $F(\varphi_2) = 2F(\varphi)$ and $F(\varphi_3) = 3F(\varphi)$ (i.e., $\varphi_3 = \psi$), then

$$3E(\varphi) - E(\psi) = k^2\sin\varphi(\sin\varphi\sin\varphi_2 + \sin\varphi_2\sin\varphi_3).$$

In order to get rid of the extra algebraic term on the right-hand side and solve the problem of the division of an arc of the ellipse into three equal parts in the strict sense, let us again construct the arc with an arbitrary beginning point. Let us confine ourselves to the division of the arc $\smile AB$ into three equal parts, where OA and OB are the major and minor semi-axes, respectively. In this case $\psi = \varphi_3 = \frac{\pi}{2}$; hence, $\sin\varphi_3 = 1$ and

$$3E(\varphi) - E(\psi) = k^2\sin\varphi\sin\varphi_2(\sin\varphi + 1),$$

i.e.,

$$E(\varphi) = \frac{1}{3}E^1 + \frac{1}{3}k^2\sin\varphi\sin\varphi_2(\sin\varphi + 1). \tag{2.2}$$

Let $x = \sin\varphi$ be a root of equation (2.1). Let us find the angles ψ and ω satisfying equations

$$E(\omega) - E(\psi) = \frac{1}{3}E^1, \tag{2.3}$$

$$F(\varphi) + F(\psi) - F(\omega) = 0. \tag{2.4}$$

From (2.4) it follows that

$$E(\varphi) + E(\psi) - E(\omega) = k^2 \sin\varphi \sin\psi \sin\omega.$$

Taking (2.2) and (2.3) into account, we get

$$\sin\psi \sin\omega = \frac{1}{3}\sin\varphi_2(1 + \sin\varphi). \tag{2.5}$$

Moreover, it follows from (2.4) that

$$\cos\psi\cos\omega + \sin\psi\sin\omega\Delta(\varphi) = \cos\varphi$$

and, therefore,

$$\cos\psi\cos\omega = \cos\varphi - \frac{1}{3}\sin\varphi_2\Delta(\varphi)(1 + \sin\varphi).$$

Making use of the relation

$$\cos\varphi_2\cos\varphi_3 + \sin\varphi_2\sin\varphi_3\Delta(\varphi) = \cos\varphi,$$

where $\varphi_3 = \frac{\pi}{2}$, we obtain $\sin\varphi_2 = \cos\varphi/\Delta(\varphi)$. Taking into account relation (2.1) with $x = \sin\varphi$, it is also possible to express this relation in the form $\cos\varphi_2 = 1 - \sin\varphi$; to this end we should express (2.1) as

$$1 - \frac{1 - x^2}{1 - k^2x^2} = (1 - x)^2.$$

Hence,

$$\cos\psi\cos\omega = \cos\varphi - \frac{1}{3}\cos\varphi(1 + \sin\varphi) = \frac{1}{3}\cos\varphi(1 + \cos\varphi_2).$$

The last relation together with (2.5) leads to the following formulas:

$$\begin{aligned}\cos(\psi \pm \omega) &= \frac{1}{3}(\cos\varphi + \cos\varphi\cos\varphi_2 \mp \sin\varphi_2 \mp \sin\varphi\sin\varphi_2)\\ &= \frac{1}{3}(\cos\varphi \mp \sin\varphi_2 + \cos(\varphi \pm \varphi_2)).\end{aligned}$$

Thus, if a root of equation (2.1) can be constructed with a ruler and compass, then for any ellipse with the ratio of its axes equal to $1 : \sqrt{1 - k^2}$ we can construct, using a ruler and compass, an arc whose length is equal to a third of the length of the arc $\smile AB$. An example of such an ellipse is an ellipse with the ratio of axes $1 : \sqrt{2}$ for which $k^2 = \frac{1}{2}$. Indeed, in this case equation (2.1) is of the form

$$x^4 - 2x^3 + 4x - 2 = 0.$$

Making the change of variables $x = \sqrt{1-z}$ we get the equation

$$2(1+z)\sqrt{1-z} = 2 - (1-z)^2.$$

Squaring both sides we get

$$z^4 + 6z^2 - 3 = 0, \quad \text{i.e.,} \quad z^2 = -3 \pm 2\sqrt{3}.$$

Finally, $z = \sqrt{2\sqrt{3} - 3}$ and $\sin\varphi = x = \sqrt{1-z}$, i.e., the angle φ can be constructed with a ruler and compass.

REMARK. The whole material of this section is borrowed from the classical treatise by Legendre [**A12**].

PROBLEMS

3.2.1. Let the angles φ and ψ be related by the formula $b\tan\varphi\tan\psi = 1$. By Fagnano's theorem $\overset{\frown}{BM} - \overset{\frown}{NA} = k^2 \sin\varphi \sin\psi$. Prove that this difference is also equal to the length of the segment MM_1 as well as to that of NN_1, where M_1 and N_1 are the projections of the center of the ellipse on the tangent lines at the points M and N (see Figure 28).

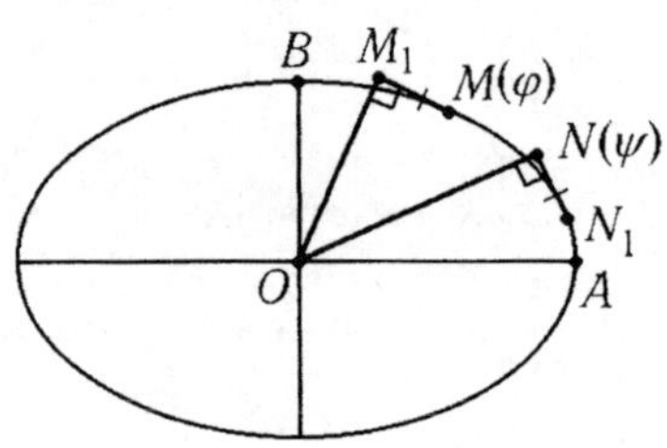

FIGURE 28

§3.3. Curves with elliptic arcs

The ellipse is an example of a curve whose arc lengths can be expressed in terms of an elliptic integral of the second kind. For a curve whose arc lengths are expressed via elliptic integrals of the first kind the addition of arc lengths can also be performed even without an algebraic extra term. Are there curves whose coordinates are sufficiently simple functions of a parameter φ and whose arc length, as a function of φ, is an elliptic integral of the first kind?

Consider the curve $x = x(\varphi)$, $y = y(\varphi)$ such that the length of the arc from the point with parameter 0 to the point with parameter φ is equal to

$$F(\varphi) = \int_0^{\varphi} \frac{d\varphi}{\sqrt{1 - k^2 \sin^2\varphi}}.$$

For $k = \frac{1}{\sqrt{2}}$ such a curve was known already to **Fagnano**; this is the lemniscate. **Legendre** attempted to construct an example of such a curve for an arbitrary k. If

$$dx = \frac{\cos\varphi\, d\varphi}{1 - k^2\sin^2\varphi} \quad \text{and} \quad dy = \frac{-b\sin\varphi\, d\varphi}{1 - k^2\sin^2\varphi},$$

where $b = \sqrt{1-k^2}$, then

$$dx^2 + dy^2 = \frac{d\varphi^2}{1 - k^2 \sin^2 \varphi} = \left(\frac{d\varphi}{\Delta(\varphi)}\right)^2, \quad \text{where} \quad \Delta(\varphi) = \sqrt{1 - k^2 \sin^2 \varphi}.$$

Therefore, as the functions $x(\varphi)$ and $y(\varphi)$ we may take primitives of the functions

$$\frac{\cos\varphi}{1 - k^2 \sin^2 \varphi} \quad \text{and} \quad \frac{-b \sin\varphi}{1 - k^2 \sin^2 \varphi}.$$

The corresponding integrals are easy to calculate using the change of variables $u = \sin\varphi$ and $v = \cos\varphi$. As a result, we get

$$x = \frac{1}{2k} \ln \frac{1 + ku}{1 - ku} = \frac{1}{2k} \ln \frac{1 + k \sin\varphi}{1 - k \sin\varphi},$$

$$y = \frac{1}{k} \arctan \frac{kv}{b} = \frac{1}{k} \arctan \frac{k \cos\varphi}{b}.$$

It is easy to verify that

$$\cos ky = \frac{b}{\sqrt{1 - k^2 \sin^2 \varphi}}$$

and

$$\cosh kx = \frac{e^{kx} + e^{-kx}}{2} = \frac{1}{\sqrt{1 - k^2 \sin^2 \varphi}},$$

i.e., $\cos ky = b \cosh kx$.

Legendre found this example easily but he was not satisfied since the curve

$$\cos ky = b \cosh kx$$

is not an algebraic one. He also succeeded in constructing an example of an algebraic curve with the arc length equal to $F(\varphi)$ (with a certain algebraic extra term).

These studies of Legendre led mathematicians to the problem of finding all algebraic curves whose arc length is an elliptic integral of the first kind. Legendre's problem was solved by the French mathematician **Joseph-Alfred Serret**. In three papers published in Liouville's Journal de l'Ecole Polytechnique he managed to construct a family of plane algebraic curves S_p depending on a positive rational parameter p whose arc length is exactly $F(\varphi)$. (For an exposition of these papers see [**A14**], §§563–565.) Moreover, Serret proved that in this way one gets all the algebraic curves with required properties.

Serret's curves are obtained in the following way. Let p be a fixed number. Consider the triangle OPM with the length of the sides OP and PM equal to $\sqrt{p}$ and $\sqrt{p+1}$, respectively, and the angles at the vertices O and M equal to α and β, respectively. Let

(3.1) $$\cos\omega = \cos(p\alpha - (p+1)\beta).$$

Let us introduce a Cartesian coordinate system Oxy with the origin at the vertex O of the triangle OPM and the angle between the x-axis and the side OM equal to ω (Figure 29).

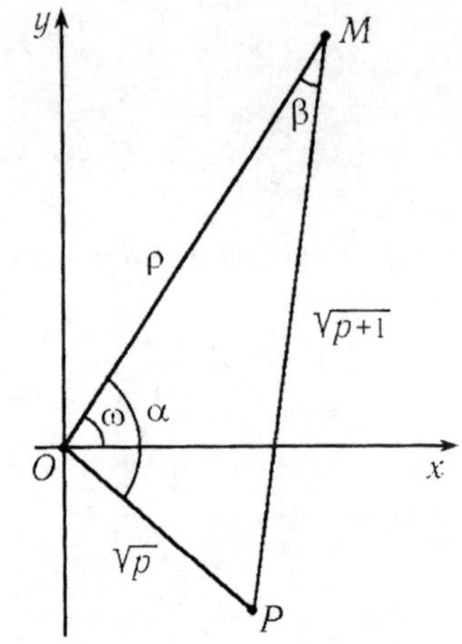

FIGURE 29

Let us vary the triangle OPM so that the point O is fixed, the lengths of the sides OP and PM are constants and the angle ω between the x-axis and the side OM is determined by relation (3.1) at all times. Then the point M will plot a certain curve S_p. Let $x = \rho\cos\omega$ and $y = \rho\sin\omega$ be the coordinates of M; we may assume that x and y depend on a parameter α.

THEOREM (Serret). *The length of an arc of S_p as a function of the parameter α is equal to*

$$\sqrt{p}\int_0^{\alpha}\frac{d\varphi}{\sqrt{1-k^2\sin^2\varphi}},\qquad \textit{where}\quad k=\sqrt{\frac{p}{p+1}}.$$

The curve S_p is algebraic for any positive rational p.

PROOF. Let $OM = \rho$. By the law of cosines $p+1 = p+\rho^2 - 2\rho\sqrt{p}\cos\alpha$ and $p = p+1+\rho^2-2\rho\sqrt{p+1}\cos\beta$, i.e.,

$$\cos\alpha = \frac{\rho^2-1}{2\rho\sqrt{p}}\quad\text{and}\quad\cos\beta=\frac{\rho^2+1}{2\rho\sqrt{p+1}}. \tag{3.2}$$

Thus, ρ and $\cos\alpha$ are related by a (polynomial) constraint $F(\cos\alpha,\rho)\equiv 0$, where $F(x,y) = 2xy\sqrt{p}-y^2+1$. Moreover, if $p\in\mathbb{Q}$, then $\cos\alpha$ and $\cos p\alpha$ are also connected by a polynomial relation. It follows that ρ and $\cos p\alpha$ are related by a polynomial dependence. Similarly, ρ and $\cos(p+1)\beta$ are related by a polynomial dependence. Since $x = \rho\cos\omega = \rho\cos(p\alpha-(p+1)\beta)$ and $y=\rho\sin\omega$, it follows that x and ρ, and also y and ρ, are related by polynomial dependencies. Therefore, x and y are related by a polynomial dependence, i.e., the curve S_p is an algebraic one.

It follows from (3.2) that

$$\sin\alpha=\frac{R}{2\rho\sqrt{p}}\quad\text{and}\quad\sin\beta=\frac{R}{2\rho\sqrt{p+1}},$$

where

$$R=\sqrt{-\rho^4+2(2p+1)\rho^2-1}.$$

Differentiating the equality $\cos\omega = \cos(p\alpha - (p+1)\beta)$ we get

$$-\sin\omega\, d\omega = -\sin(p\alpha - (p+1)\beta)(p\, d\alpha - (p+1)\, d\beta).$$

Since $\sin\omega = \pm\sin(p\alpha - (p+1)\beta)$, it follows that

$$\pm d\omega = p d\alpha - (p+1) d\beta.$$

Making use of the fact that

$$\cos\alpha = \frac{\rho^2-1}{2\rho\sqrt{p}} \quad \text{and} \quad d\cos\alpha = -\sin\alpha d\alpha = -\frac{R}{2\rho\sqrt{p}} d\alpha$$

we get $d\alpha = -\frac{\rho^2+1}{R}\frac{d\rho}{\rho}$. Similarly, $d\beta = -\frac{\rho^2-1}{R}\frac{d\rho}{\rho}$. Therefore,

$$\pm d\omega = p d\alpha - (p+1) d\beta = \frac{\rho^2 - (2p+1)}{R}\frac{d\rho}{\rho}.$$

Let dl be the differential of the arc length. Then

$$dl^2 = d\rho^2 + \rho^2 d\omega^2 = 4p(p+1)\frac{d\rho^2}{R^2},$$

i.e., $dl = \pm 2\sqrt{p(p+1)}\frac{d\rho}{R}$. Since $\frac{d\alpha}{\cos\beta} = -2\sqrt{p+1}\frac{d\rho}{R}$, it follows that $\pm dl = \sqrt{p}\frac{d\alpha}{\cos\beta}$. Moreover, $\sin\beta = \sqrt{\frac{p}{p+1}}\sin\alpha$; hence, $\cos\beta = \sqrt{1-k^2\sin^2\alpha}$, where $k = \sqrt{\frac{p}{p+1}}$. Finally, up to a sign, we get

$$dl = \sqrt{p}\frac{d\alpha}{\sqrt{1-k^2\sin^2\alpha}},$$

as was required. □

The curve S_p possesses the following remarkable property. Let the points O, A, and B correspond to the values of the parameters 0, α, and β, respectively. Consider a point C such that $\overset{\frown}{OC} = \overset{\frown}{OA} + \overset{\frown}{OB}$. In other words, C corresponds to the value of the parameter γ such that $F(\gamma) = F(\alpha) + F(\beta)$. Then $\cos\gamma$ can be algebraically expressed in terms of $\cos\alpha$ and $\cos\beta$. Moreover, it is also clear that the coordinates of the points A, B, and C can be algebraically expressed in terms of $\cos\alpha$, $\cos\beta$, and $\cos\gamma$, respectively. Hence, the coordinates of C can be algebraically expressed in terms of the coordinates of A and B. Thus, the addition and the division of arc lengths on the curve S_p are algebraic problems. In particular, we may write an equation for the division of an arc (with the beginning point at O) of the curve S_p into n parts of equal length.

Now, let us consider in more detail the curve S_p for $p = 1$. In this case

$$\cos\alpha = \frac{\rho^2-1}{2\rho} \quad \text{and} \quad \cos\beta = \frac{\rho^2+1}{2\sqrt{2}\rho}.$$

Easy calculations show that

$$x = \rho\cos(\alpha - 2\beta) = \frac{\rho^4-1}{4\rho^2} + 1 \quad \text{and} \quad y = \frac{R(1-\rho^2)}{4\rho^2},$$

where $R = \sqrt{-\rho^4 + 6\rho^2 - 1}$; hence, $x - 1 = \rho_1 \cos\beta$ and $y = -\rho_1 \sin\beta$, where $\rho_1 = \frac{\rho^2 - 1}{\sqrt{2}\rho}$. Moreover,

$$\cos 2\beta = \left(\frac{\rho^2 - 1}{2\rho}\right)^2 = \frac{1}{2}\rho_1^2,$$

i.e., the curve is the lemniscate.

§3.4. Curves whose arc lengths can be expressed in terms of arc lengths of the circle

In the preceding section we constructed a series of curves for which the addition of arc lengths is an algebraic operation. There are other examples of such curves. It was **Euler** who already constructed them. He wrote that he had found these curves only after long study. The coordinates of points on these curves are algebraic functions of the argument $\tan s$, where s is a parameter proportional to the length of the arc of the curve. The addition of arc lengths of such curves is an algebraic operation since $\tan(\alpha + \beta)$ can be algebraically expressed in terms of $\tan\alpha$ and $\tan\beta$.

Later on, **Serret** generalized the examples constructed by Euler. Serret even gave a complete classification of such curves. We will confine ourselves to the simplest example.

Let $x + iy = \frac{(t-a)^{n+2}}{(t-\alpha)^n (t+i)^2}$, where n is an integer or a rational number, and a and α are complex numbers such that $\alpha = \bar{a}$. As t varies from $-\infty$ to $+\infty$, the points with coordinates (x, y) form a curve. This curve is an algebraic one, because x and y were algebraically expressed in terms of t.

It is easy to verify that

$$dx + idy = \frac{(t-a)^{n+1}(pt+q)}{(t-\alpha)^{n+1}(t+i)^3}\,dt.$$

It is possible to select numbers a and $\alpha = \bar{a}$ so that $pt + q = k(t - i)$. We can even assume that $|a| = 1$, namely,

$$a = \frac{\sqrt{n(n+2)}}{n+1} - \frac{i}{n+1}.$$

For this value of a we get $pt + q = k(t - i)$, where

$$k = \frac{2\sqrt{n(n+2)}}{n+1}.$$

In this case

$$dx - idy = k\frac{(t-\alpha)^{n+1}(t+i)}{(t-a)^{n+1}(t-i)^3}\,dt.$$

Therefore,

$$dl^2 = dx^2 + dy^2 = k^2 \frac{dt^2}{(t+i)^2(t-i)^2} = \left(k\frac{dt}{t^2+1}\right)^2.$$

Hence,

$$l = \pm k \int_0^t \frac{d\tau}{\tau^2 + 1} = \pm k \arctan t.$$

Thus, if the parameter s is equal to $\frac{1}{k}$ of the arc length of the curve, then $t = \pm\tan s$ and the coordinates x and y of the points of the curve can be algebraically expressed in terms of t. The lengths of arcs are measured from the point corresponding to the parameter $t = 0$.

Problems

3.4.1. Prove that for $n = 1$ the equation of the curve considered above can be expressed (after a homothetic transformation) in polar coordinates in the form

$$\cos\varphi = \frac{\rho^2 + 6\rho - 2}{3\rho^2\sqrt{3}}.$$

CHAPTER 4

Abel's Theorem on Division of Lemniscate

A *lemniscate* is a curve whose equation in the polar coordinates is of the form $r^2 = \cos 2\theta$ (Figure 30). The name "lemniscate" stems from the Latin word *lemniscatus* — decorated with ribbons. In Cartesian coordinates (x, y), where $x = r\cos\theta$, $y = r\sin\theta$, the equation of this curve is

$$(x^2 + y^2)^2 = x^2 - y^2.$$

Indeed, $x^2 + y^2 = r^2$ and $x^2 - y^2 = r^2 \cos 2\theta$.

FIGURE 30

The French astronomer of Italian origin Jean-Dominique (Giovanni Dominico) **Cassini** (1625–1712) was the first to study the lemniscate. He considered even more general curves for whose points the product of the distances to the two fixed points F_1 and F_2 is a constant (Figure 31).

FIGURE 31

On the basis of astronomical observations Cassini believed that with the help of such curves the movement of planets could be described more precisely than with the help of ellipses. Now these curves are called *Cassini's ovals*. But Cassini's book *Eléments d'astronomie* (*Foundations of Astronomy*), in which they were studied, was published in 1749, many years after his death. To the mathematical community the lemniscate became known through papers by **J. Bernoulli** and **I. Bernoulli** published in 1694 and, therefore, it is usually called *Bernoulli's lemniscate*.

The most remarkable properties of the lemniscate were discovered by an Italian mathematician Count **Fagnano** (1682–1766). By the way, it was Fagnano who coined the term *elliptic integrals*. Fagnano discovered that the arc length of the lemniscate can be expressed in terms of an elliptic integral of the first kind. He obtained an addition theorem for this integral and, therefore, demonstrated that the division of arcs of the lemniscate into n equal parts is an algebraic problem. In 1750 Fagnano published a collection of his papers under the name *Produzioni matematiche*. The Berlin Academy asked **Leonard Euler** to write a review of this book. Fagnano's works stimulated Euler's interest in elliptic integrals. In his numerous studies Euler considerably developed and generalized Fagnano's methods and results.

Already Fagnano knew that the division of the lemniscate could be reduced to a solution of an algebraic equation. However, methods for investigating solvability of equations in quadratures (i.e., via square roots) were not yet developed at that time. The first to achieve essential progress in this field was 19 year-old **Gauss** (in 1796). He found that a regular 17-gon can be constructed with a ruler and compass, i.e., the equation $x^{17} - 1 = 0$ is solvable in square roots. Later, Gauss showed that with a ruler and compass one can construct a regular n-gon for any n of the form $2^a p_1 \cdots p_k$, where p_i are distinct *Fermat primes*, i.e., the primes of the form $2^{2^m} + 1$. Gauss wrote that for all other n it is impossible to construct a regular n-gon with a ruler and compass but we have no substantiation of the claim that he could actually prove this.

Gauss was also interested in the equation for the division of the lemniscate. For example, he showed that an equation of 25th degree related to the division of the lemniscate into 5 equal parts is solvable in square roots. His arguments were based on the fact that 5 can be represented as the product of $2+i$ by $2-i$ (for details see §4.3). Gauss did not publish these investigations but in his book *Disquisitiones Arithmeticae* (*Arithmetical Studies*), which appeared in 1801, he mentioned that the methods he developed were applicable not only to trigonometric functions but also to the functions related to the integrals of the form $\int \frac{dx}{\sqrt{1-x^4}}$.

This claim intrigued **Abel**. Abel investigated in detail the equation for the division of the lemniscate and proved that the lemniscate can be divided into n equal parts for all numbers n of the form $2^a p_1 \cdots p_k$, where the p_i are distinct Fermat primes. Abel considered this theorem as one of his most important results. It is contained in the second part of his large work *Recherches sur les fonctions elliptiques* (*Studies on Elliptic Functions*). Abel's proof is quite long and complicated.

Later, **Eisenstein** (1823–1852) obtained a simpler proof. In doing so he discovered certain interesting properties of polynomials related to the division of the lemniscate. Eisenstein's proof is still quite cumbersome.

Relatively recently **Rosen** (see [**C14**]) found an elegant proof of Abel's theorem. His proof is rather simple and, moreover, it distinctly demonstrates the decisive role played by the invariance of the period lattice of lemniscatic functions with respect to multiplication by i.

To help the reader to feel the style of Abel's era and, at the same time, become acquainted with the modern interpretation of the subject, we will reproduce both proofs: the one that Eisenstein obtained one and a half centuries ago and the one recently found by Rosen.

Abel only proved the possibility of the division of the lemniscate into n equal parts with a ruler and compass for the indicated values of n. He did not prove that for the other values of n this is impossible. In [**C14**] it is shown that for the other values of n it is impossible to construct the coordinates of the points that divide the lemniscate into n equal parts with a ruler and compass. This, however, does not mean that for the other values of n it is impossible to divide the lemniscate into n equal parts with a ruler and compass if the lemniscate *is already drawn*. Indeed, use of the lemniscate *itself* provides us with additional possibilities for constructions. Considering the points of intersection of the straight lines and circles with the lemniscate one can, in general, construct more than just quadratic irrationalities.

* * *

Before we plunge into the study of the equation for the division of the lemniscate, let us consider a simpler equation for the division of the circle. First, we will show how to solve in square roots the equation $x^{17} - 1 = 0$ by a quite elementary method, though this solution cannot be generalized to the equation for the division of the lemniscate. Next, we will discuss the approach to the study of the solvability of the equation $x^n - 1 = 0$ in square roots that can be generalized to the equation for the division of the lemniscate.

§4.1. Construction of a regular 17-gon. An elementary approach

The roots of the equation $x^n - 1 = 0$ are the vertices of a regular n-gon. Indeed, if $\varepsilon = \exp(2\pi i/n)$, then $\varepsilon, \varepsilon^2, \dots, \varepsilon^n = 1$ are the roots of this equation. Dividing the polynomial $x^n - 1$ by $x - 1$ we get the polynomial $x^{n-1} + x^{n-2} + \dots + x + 1$. Thus, if the equation

$$x^{n-1} + x^{n-2} + \dots + x + 1 = 0 \tag{1.1}$$

is solvable in square roots, then it is possible to construct a regular n-gon with a ruler and compass.

For $n = 3$ there is no problem, since the quadratic $x^2 + x + 1 = 0$ is, without doubt, solvable in square roots. For $n = 5$ equation (1.1) is also easy to solve. Indeed, the substitution $u = x + x^{-1}$ turns it into $u^2 + u - 1 = 0$.

For $n = 17$ it is not that easy to solve equation (1.1) in square roots. To do so, Gauss used a special partition of the numbers ε, ε^2, ε^3, ... , ε^{16} into groups, where $\varepsilon = \exp(2\pi i/17)$. To get such a partition, we enumerate the given numbers so that for a fixed l the root ε_{k+l} is obtained from ε_k in the same fashion, namely, by raising to a fixed power: $\varepsilon_{k+l} = (\varepsilon_k)^c$:

$$\varepsilon_k \varepsilon_l = \varepsilon_{k+l}.$$

Such a numeration can be obtained by setting $\varepsilon_k = \varepsilon^{g^k}$, where the residues of the numbers 1, g, g^2, ... , g^{15} after the division by 17 take all values from 1 to 16. It is easy to see that $g = 3$ possesses this property. For $g = 3$ the numbers $\varepsilon_0, \dots, \varepsilon_{15}$ and their respective values are written one under another in the following table:

ε	ε^3	ε^9	ε^{10}	ε^{13}	ε^5	ε^{15}	ε^{11}	ε^{16}	ε^{14}	ε^8	ε^7	ε^4	ε^{12}	ε^2	ε^6
ε_0	ε_1	ε_2	ε_3	ε_4	ε_5	ε_6	ε_7	ε_8	ε_9	ε_{10}	ε_{11}	ε_{12}	ε_{13}	ε_{14}	ε_{15}

Let x_1 be the sum of the numbers ε_k with even indices k, and x_2 the sum of the numbers ε_k with odd indices k, i.e.,

$$x_1 = \varepsilon + \varepsilon^9 + \varepsilon^{13} + \varepsilon^{15} + \varepsilon^{16} + \varepsilon^8 + \varepsilon^4 + \varepsilon^2,$$

$$x_2 = \varepsilon^3 + \varepsilon^{10} + \varepsilon^5 + \varepsilon^{11} + \varepsilon^{14} + \varepsilon^7 + \varepsilon^{12} + \varepsilon^6.$$

The sum of all the roots of the equation $x^{17} - 1 = 0$ (the root $x = 1$ included) is equal to zero, hence, $x_1 + x_2 = -1$. Simple calculations show that $x_1 x_2 = -4$. Indeed, let $\alpha = 2\pi/17$. Then $\varepsilon^k = \cos k\alpha + i \sin k\alpha$; hence,

$$\begin{aligned} \varepsilon + \varepsilon^{16} &= 2\cos\alpha, & \varepsilon^9 + \varepsilon^8 &= 2\cos 8\alpha, \\ \varepsilon^{13} + \varepsilon^4 &= 2\cos 4\alpha, & \varepsilon^{15} + \varepsilon^2 &= 2\cos 2\alpha, \end{aligned}$$

i.e.,

$$x_1 = 2(\cos\alpha + \cos 8\alpha + \cos 4\alpha + \cos 2\alpha).$$

Similarly,

$$x_2 = 2(\cos 3\alpha + \cos 7\alpha + \cos 5\alpha + \cos 6\alpha).$$

Using the formula

$$2\cos p\alpha \cos q\alpha = \cos(p+q)\alpha + \cos(p-q)\alpha$$

we get

$$x_1 x_2 = 8(\cos\alpha + \cos 2\alpha + \cos 3\alpha + \dots + \cos 8\alpha) = 4(x_1 + x_2) = -4.$$

Thus, we can find x_1 and x_2 from the quadratic equation

$$x^2 + x - 4 = 0. \tag{1.2}$$

Since

$$\cos\alpha + \cos 2\alpha > 2\cos\frac{\pi}{4} = \sqrt{2} > -\cos 8\alpha$$

and $\cos 4\alpha > 0$, it follows that $x_1 > 0$. Hence, $x_2 = -\frac{4}{x_1} < 0$, i.e., x_1 is the positive root of equation (1.2) and x_2 is the negative root.

Denoting by y_1, y_3, y_2 and y_4 the sums of the numbers ε_k with indices whose residues modulo 4 are equal to 0, 1, 2 and 3, respectively, we get

$$\begin{aligned} y_1 &= \varepsilon + \varepsilon^{13} + \varepsilon^{16} + \varepsilon^4 = 2(\cos\alpha + \cos 4\alpha), \\ y_2 &= \varepsilon^9 + \varepsilon^{15} + \varepsilon^8 + \varepsilon^2 = 2(\cos 8\alpha + \cos 2\alpha), \\ y_3 &= \varepsilon^3 + \varepsilon^5 + \varepsilon^{14} + \varepsilon^{12} = 2(\cos 3\alpha + \cos 5\alpha), \\ y_4 &= \varepsilon^{10} + \varepsilon^{11} + \varepsilon^7 + \varepsilon^6 = 2(\cos 7\alpha + \cos 6\alpha). \end{aligned}$$

It is clear that $y_1 + y_2 = x_1$ and $y_1 > y_2$, because $\cos\alpha > \cos 2\alpha$ and $\cos 4\alpha > \cos 8\alpha$. Moreover,

$$y_1 y_2 = 4(\cos\alpha + \cos 4\alpha)(\cos 8\alpha + \cos 2\alpha) = 2(\cos\alpha + \dots + \cos 8\alpha) = -1.$$

Therefore, y_1 and y_2 satisfy the equation $y^2 - x_1 y - 1 = 0$. It is easy to verify that y_3 and y_4 satisfy the equation $y^2 - x_2 y - 1 = 0$; moreover, $y_3 > y_4$.

Finally, let us consider $z_1 = \varepsilon + \varepsilon^{16} = 2\cos\alpha$ and $z_2 = \varepsilon^{13} + \varepsilon^4 = 2\cos 4\alpha$, i.e., the sums of numbers ε_k with indices whose residues after the division by 8 are equal to 0 and 4, respectively. Then $z_1 > z_2$, $z_1 + z_2 = y_1$ and

$$z_1 z_2 = 4\cos\alpha\cos 4\alpha = 2(\cos 5\alpha + \cos 3\alpha) = y_3.$$

Therefore, z_1 is the largest root of the equation $z^2 - y_1 z + y_3 = 0$. Thus, the segment of length $z_1 = 2\cos(2\pi/17)$ can be constructed with a ruler and compass. Now it is clear how to construct a regular 17-gon.

§4.2. Construction of regular polygons. Elements of Galois theory

In the preceding section we showed how to solve in square roots the equation $x^{17} - 1 = 0$. Now we prove that for all numbers n of the form $2^n p_1 \cdots p_k$, where the p_i are distinct Fermat primes, the equation $x^n - 1 = 0$ is also solvable in square roots. Our exposition will be such that a good deal of it can be generalized to the case of the lemniscate almost without changes.

Assigning to every real number t the point with coordinates $(\cos t, \sin t)$, we get a parameterization of the unit circle C by real numbers. As a result, C turns into an abelian group with unit element $(1, 0)$.

Since

$$\cos(t+s) = \cos t\cos s - \sin t\sin s \quad\text{and}\quad \sin(t+s) = \sin t\cos s + \cos t\sin s,$$

the law of addition of points on this circle can be expressed as follows:

$$(a,b) + (c,d) = (ac - bd, ad + bc) = (f(a,b,c,d), g(a,b,c,d)).$$

It is easy to verify that

$$2(x,y) = (x,y) + (x,y) = (x^2 - y^2, 2xy)$$

and

$$3(x,y) = (x^3 - 3xy^2, 3x^2y - y^3).$$

Similarly,

$$n(x,y) = (f_n(x,y), g_n(x,y)),$$

where f_n and g_n are polynomials with integer coefficients. From the relation $\cos n\varphi + i\sin n\varphi = (\cos\varphi + i\sin\varphi)^n$ we get

$$f_n(x,y) = \frac{(x+iy)^n + (x-iy)^n}{2}, \quad g_n(x,y) = \frac{(x+iy)^n - (x-iy)^n}{2i}. \tag{2.1}$$

Let C_n be the set of points $(x,y) \in C$ such that $n(x,y) = (1,0)$, i.e., $f_n(x,y) = 1$ and $g_n(x,y) = 0$. These points can serve as vertices of a regular n-gon. It is also clear that C_n is a subgroup of C isomorphic to $\mathbb{Z}/n\mathbb{Z}$, the additive group of residues modulo n.

Over $\mathbb{C}$, in addition to the points of C_n there are other solutions of the system

$$f_n(x,y) = 1, \quad g_n(x,y) = 0.$$

Let us find all these solutions. Using formulas (2.1) we can pass to an equivalent system of equations

$$(x+iy)^n = 1, \quad (x-iy)^n = 1.$$

Therefore, $x+iy=\varepsilon^p$ and $x-iy=\varepsilon^q$ for $\varepsilon=\exp(2\pi i/n)$. In particular, $x^2+y^2=\varepsilon^p\varepsilon^q$; hence, the equality $x^2+y^2=1$ holds if and only if $\varepsilon^p=\varepsilon^{-q}$. Thus, C_n can be characterized as the set of all solutions of the system of equations

$$f_n(x,y)=1, \quad g_n(x,y)=0, \quad x^2+y^2=1. \tag{2.2}$$

Let us consider the field K_n generated over $\mathbb{Q}$ by the Cartesian coordinates of all the points of C_n. For example, C_3 consists of points (1, 0) and $(-\frac{1}{2}, \pm\frac{\sqrt{3}}{2})$; and, therefore, $K_3=\mathbb{Q}(\sqrt{3})$; C_4 consists of the points $(\pm 1, 0)$ and $(0, \pm 1)$; hence, $K_4=\mathbb{Q}$.

Let σ be an automorphism of K_n identical on $\mathbb{Q}$. Since the coefficients of the polynomials f_n and g_n are integers, σ sends any solution of system (2.2) in another solution of this system, i.e., σ determines a permutation of points of C_n. The automorphism σ can be uniquely recovered from this permutation because the coordinates of points of C_n generate the field K_n. It is also clear that

$$\sigma\left((a,b)+(c,d)\right)=\sigma(ac-bd, ad+bc)=\sigma(a,b)+\sigma(c,d),$$

i.e., it is not an arbitrary permutation of points of C_n that corresponds to σ but an automorphism of the group $\mathbb{Z}/n\mathbb{Z}$. Therefore, the group G_n of automorphisms of K_n over $\mathbb{Q}$ is isomorphic to a subgroup of the group $\operatorname{Aut}(\mathbb{Z}/n\mathbb{Z})$.

The idea of the remaining part of the proof is as follows. First, we will show that if $n=2^a p_1\cdots p_k$, where the p_i are distinct Fermat primes, then the order of the group $\operatorname{Aut}(\mathbb{Z}/n\mathbb{Z})$ is a power of 2. In particular, the order of G_n is also a power of 2.

Next, we will prove that if the order of the group G is equal to 2^k, then there exists a sequence of subgroups

$$G=G^0\supset G^1\supset\cdots\supset G^k=\{e\}$$

such that G^i is a subgroup of G^{i-1} of index 2 for $i=1, \dots, k$.

Finally, using this sequence of subgroups we will construct a sequence of quadratic field extensions beginning with $\mathbb{Q}$ and terminating with K_n. The existence of such a sequence of extensions implies that all elements of K_n (in particular, the coordinates of points of C_n) are quadratic irrationalities, i.e., can be constructed using a sequence of square roots.

4.2.1. Lemma. *The order of the group* $\operatorname{Aut}(\mathbb{Z}/n\mathbb{Z})$ *is equal to a power of* 2 *if and only if* $n=2^a p_1\cdots p_k$, *where the* p_i *are distinct Fermat primes.*

Proof. As is known, all automorphisms of the additive group $\mathbb{Z}/n\mathbb{Z}$ are of the form $x\longmapsto mx$, where m is a number relatively prime with n. Therefore, the order of $\operatorname{Aut}(\mathbb{Z}/n\mathbb{Z})$ is equal to $\varphi(n)$, where $\varphi(n)$ is the number of positive integers that do not exceed n and are relatively prime with n. If numbers p and q are relatively prime, then $\varphi(pq)=\varphi(p)\varphi(q)$. Indeed, let $0\le a\le p-1$ and $0\le b\le q-1$. Then *the residues after the division of* $n=pq$ *numbers of the form* $aq+bp$ *by* n *form the complete system of residues modulo* n.

To prove this, it suffices to observe that $a_1q+b_1p\equiv a_2q+b_2p \pmod{pq}$ if and only if $(a_1-a_2)q\equiv(b_2-b_1)p \pmod{pq}$, i.e., $a_1\equiv a_2 \pmod{p}$ and $b_1\equiv b_2 \pmod{q}$. It is also clear that the numbers $aq+bp$ and pq are relatively prime if and only if a and p and also b and q are relatively prime.

If $n = p^k$, where p is a prime, then $\varphi(n) = p^{k-1}(p-1)$. Indeed, among the numbers that do not exceed n only the numbers p, $2p$, ..., $p^{k-1}p$ have a common divisor with n. The total number of such numbers is equal to p^{k-1}.

Let $n = p_1^{k_1} \cdots p_m^{k_m}$. Then $\varphi(n)$ is the product of the numbers $p_i^{k_i-1}(p_i - 1)$. The number $p^{k-1}(p-1)$ can be a power of 2 only in the following two cases:

a) $p = 2$ and k is an arbitrary positive integer;

b) $p - 1 = 2^c$ and $k = 1$.

In the second case the number c cannot have odd divisors. Indeed, if d is an odd divisor of c, then $2^c + 1 = x^d + 1$ is divisible by $x + 1$. Hence, p is a prime of the form $2^{2^l} + 1$. □

4.2.2. LEMMA. *If the order of a group G is equal to 2^k, then there exists a sequence of subgroups*

$$G = G^0 \supset G^1 \supset \cdots \supset G^k = \{e\}$$

such that G^i is a subgroup of G^{i-1} of index 2 for $i = 1, \ldots, k$.

PROOF. Let us apply induction on k. For $k = 1$ the statement is obvious. Suppose that the statement is proved for all groups of order 2^{k-1}. For every element $x \in G$ consider the class of elements conjugate to it, i.e., the set $[gxg^{-1}]$ of elements of the form gxg^{-1}, where $g \in G$. Any two such classes either coincide or are disjoint, i.e., G is divided into the union of nonintersecting classes of conjugate elements. The equality $g_1xg_1^{-1} = g_2xg_2^{-1}$ is equivalent to the equality $xh = hx$, where $h = g_2^{-1}g_1$. Consider the subgroup

$$G_x = \{h \in G \mid xh = hx\}.$$

The elements $g_1xg_1^{-1}$ and $g_2xg_2^{-1}$ are equal if and only if $g_1 \in g_2G_x$. Therefore, the number of elements in the class $[gxg^{-1}]$ is equal to the index of the subgroup G_x in G; hence, it is of the form 2^s.

The class $[gxg^{-1}]$ contains exactly one element only if x commutes with all elements of G, i.e., x is an element of the center of G. Suppose that the center of G consists of the unit element only. Then the sum of cardinalities of all the conjugacy classes is equal to $1 + 2^{s_1} + \cdots + 2^{s_p}$, where $s_i \geq 1$. Hence, this sum is an odd number. On the other hand, it is equal to the order of G, i.e., it is equal to 2^k. This contradiction implies that the center of G contains an element $a \neq e$.

The element a generates a cyclic subgroup of order 2^r. Let us consider an element $b = a^m$, where $m = 2^{r-1}$. The element b generates a subgroup H of order 2 that belongs to the center of G; in particular, H is a normal subgroup. By the induction hypothesis for the group $F = G/H$ of order 2^{k-1}, there exists a sequence

$$F = F^0 \supset F^1 \supset \cdots \supset F^{k-1} = \{e\},$$

where F^i is a subgroup of F^{i-1} of index 2 for $i = 1, \ldots, k-1$. To get the required sequence of subgroups, set $G^k = H$ and $G^i = F^i \cup bF^i$ for $i = 0, \ldots, k-1$. □

Let $n = 2^a p_1 \cdots p_m$, where the p_i are distinct Fermat primes. Then the group G_n of automorphisms of the field K_n over $\mathbb{Q}$ has a sequence of subgroups

$$G_n = G^0 \supset G^1 \supset \cdots \supset G^k = \{e\},$$

where G^i is a subgroup of G^{i-1} of index 2 for $i = 1, \ldots, k$. To the subgroup G^i we assign the set L^i consisting of the elements of K_n that are fixed under the action of all the automorphisms in G^i. Since the sum, difference, product, and ratio of elements from L^i belong to L^i, it follows that L^i is a field. It is clear that $L^k = K_n$ and $L^i \supset L^{i-1}$.

We can show that $L^0 = \mathbb{Q}$, i.e., for any $x \in K_n \backslash \mathbb{Q}$ there exists an automorphism of the field K_n over $\mathbb{Q}$ that moves x. In general, not every extension of $\mathbb{Q}$ possesses such a property. For example, the field

$$\{p + q\sqrt[3]{2} + r\sqrt[3]{4} \mid p, q, r \in \mathbb{Q}\}$$

has no automorphisms distinct from the identity one. Indeed, the element $\sqrt[3]{2}$ can only pass into a root of the equation $x^3 - 2 = 0$, but only one root of this equation belongs to the field considered.

The reasons why $L^0 = \mathbb{Q}$ will be given a little later. For the moment let us assume this without proof.

All automorphisms in G^i preserve the elements of L^i. Moreover, $G^{i-1} = G^i \cup \sigma G^i$, where $\sigma \in G^{i-1}$. Hence, $\sigma^2 \in G^i$ because σ^2 cannot belong to σG^i. Therefore, the automorphism σ of K_n is such that if $x \in L^i$, then $\sigma^2 x = x$. Moreover, $\sigma x = x$ if and only if $x \in L^{i-1}$. Any element $x \in L^i$ can be represented as the sum of elements $x_1 = \frac{1}{2}(x + \sigma x)$ and $x_2 = \frac{1}{2}(x - \sigma x)$, where $\sigma x_1 = x_1$ and $\sigma x_2 = -x_2$. Therefore, $x_1 \in L^{i-1} \subset L^i$ and $x_2 = x - x_1 \in L^i$.

Suppose that $L^i \neq L^{i-1}$. Let $a \in L^i \setminus L^{i-1}$ and $\alpha = \sigma a - a$. Then $\sigma\alpha = -\alpha$, where $\alpha \neq 0$. Moreover, $\sigma(\alpha x_2) = (-\alpha)(-x_2) = \alpha x_2$, i.e., $\alpha x_2 \in L^{i-1}$ and $x_2 \in \alpha^{-1}L^{i-1}$. Therefore, $L^i = L^{i-1} + \alpha^{-1}L^{i-1}$. Hence, if $x \in L^i$, then the elements 1, x, x^2 are linearly dependent over L^{i-1}; therefore, $x^2 + px + q = 0$, where $p, q \in L^{i-1}$. It follows that any element from K_n is a quadratic irrationality over L^0.

Now let us prove that $L^0 = \mathbb{Q}$. First, observe that K_n is an algebraic extension of $\mathbb{Q}$, i.e., the coordinates of points of C_n are algebraic numbers. Indeed, the solutions of the system of equations

$$f_n(x, y) = 1, \quad g_n(x, y) = 0$$

are algebraic; in order to prove this, it suffices to consider $f_n(x, y) - 1$ and $g_n(x, y)$ as polynomials of y and examine their resultant.

Let $\overline{\mathbb{Q}}$ be the set of all algebraic numbers (over $\mathbb{Q}$). It is easy to verify that $\overline{\mathbb{Q}}$ is a field. Indeed, let $\alpha, \beta \in \overline{\mathbb{Q}}$. Then $\alpha^p = a_0 + a_1\alpha + \cdots + a_{p-1}\alpha^{p-1}$ and $\beta^q = b_0 + b_1\beta + \cdots + b_{q-1}\beta^{q-1}$, where $a_i, b_j \in \mathbb{Q}$ and $a_0 b_0 \neq 0$. Therefore, any element of the ring generated over $\mathbb{Q}$ by elements α and β can be represented in the form of a linear combination with rational coefficients of elements $\alpha^i\beta^j$, where $0 \leq i < p$ and $0 \leq j < q$. In particular, each of the elements $1, \alpha+\beta, \ldots, (\alpha+\beta)^{pq}$ can be represented as a linear combination of the pq elements indicated; hence, they are linearly dependent over $\mathbb{Q}$, i.e., $\alpha + \beta \in \overline{\mathbb{Q}}$.

We similarly prove that $\alpha\beta \in \overline{\mathbb{Q}}$. Moreover, $a_0\alpha^{-1} = \alpha^{p-1} - a_1 - \cdots - a_{p-1}\alpha^{p-2}$; hence, $\alpha^{-1} \in \overline{\mathbb{Q}}$.

Any automorphism τ of the field $\overline{\mathbb{Q}}$ over $\mathbb{Q}$ sends K_n into itself. Indeed, the field K_n is generated by coordinates of points of C_n and C_n coincides with the set of all solutions (over $\mathbb{C}$) of the system of equations

$$f_n(x, y) = 1, \quad g_n(x, y) = 0, \quad x^2 + y^2 = 1.$$

The equality $L^0 = \mathbb{Q}$ means that

if $a \in K_n \setminus \mathbb{Q}$, *then there exists an automorphism* $\sigma : K_n \longrightarrow K_n$ *for which* $\sigma(a) \neq a$.

To prove this, it suffices to indicate an automorphism $\tau : \overline{\mathbb{Q}} \longrightarrow \overline{\mathbb{Q}}$ for which $\tau(a) \neq a$.

4.2.3. THEOREM. *Let* a *and* b *be two roots of an irreducible polynomial over* $\mathbb{Q}$. *Then there exists an automorphism* $\tau : \overline{\mathbb{Q}} \longrightarrow \overline{\mathbb{Q}}$ *such that* $\tau(a) = b$.

PROOF. Let K be a field, α a root of an irreducible polynomial P over K, and $K(\alpha)$ the field generated by α over K. Then an arbitrary isomorphism $f : K \longrightarrow K'$ can be extended to an isomorphism $g : K(\alpha) \longrightarrow K'(\beta)$, where β is a root of the polynomial $f(P)$. Indeed, the field $K(\alpha)$ consists of the elements of the form $\sum k_j\alpha^j$, where $j \geq 0$ and $k_j \in K$. Set $g(\sum k_j\alpha^j) = \sum f(k_j)\beta^j$. This map is well defined because the equality $\sum k_j\alpha^j = 0$ is equivalent to the fact that the polynomial $F = \sum k_j x^j$ is divisible by P.

Let us first construct an isomorphism $\tau_1 : \mathbb{Q}(a) \longrightarrow \mathbb{Q}(b)$. Then select an element $t_2 \in \overline{\mathbb{Q}} \setminus \mathbb{Q}(a)$. It is a root of an irreducible polynomial P_2 over $\mathbb{Q}(a)$. Let t_2' be a root of the polynomial $\tau_1(P_2)$.

Next, we can construct an isomorphism $\tau_2 : \mathbb{Q}(a, t_2) \longrightarrow \mathbb{Q}(b, t_2')$. Select an element $t_3 \in \overline{\mathbb{Q}} \setminus \mathbb{Q}(a, t_2)$ and construct an isomorphism $\tau_3 : \mathbb{Q}(a, t_2, t_3) \longrightarrow \mathbb{Q}(b, t_2', t_3')$, etc. Since the dimension of $\overline{\mathbb{Q}}$ over $\mathbb{Q}$ is countable, we can construct a basis $\{1, \varepsilon_1 = a, \varepsilon_2, \varepsilon_3, \dots\}$ of $\overline{\mathbb{Q}}$ over $\mathbb{Q}$. The elements t_2, t_3, ... can be chosen so that the field $\mathbb{Q}(a, t_2, \dots, t_k)$ contains a subspace generated by elements $1, \varepsilon_1, \dots, \varepsilon_k$.

As a result, we get a monomorphism $\tau : \overline{\mathbb{Q}} \longrightarrow \overline{\mathbb{Q}}$ such that $\tau(a) = b$. It remains to verify that τ is an epimorphism. Let $\gamma_1 \in \overline{\mathbb{Q}}$ be a root of an irreducible polynomial R over $\mathbb{Q}$ and let $\gamma_1, \dots, \gamma_n$ be all the roots of this polynomial. Then $\tau(\gamma_i) \in \{\gamma_1, \dots, \gamma_n\}$, where all the numbers $\tau(\gamma_1), \dots, \tau(\gamma_n)$ are distinct. Hence, the sets $\{\gamma_1, \dots, \gamma_n\}$ and $\{\tau(\gamma_1), \dots, \tau(\gamma_n)\}$ coincide. In particular, $\gamma_1 = \tau(\gamma_j)$ for some j. □

It is possible to significantly generalize Theorem 4.2.3. First observe that $\overline{\mathbb{Q}}$ is algebraically closed. Indeed, let x_0 be a root of the polynomial $\alpha + \beta x + \dots + \omega x^n = 0$, where $\alpha, \beta, \dots, \omega \in \overline{\mathbb{Q}}$. Let us consider a polynomial R irreducible over $\mathbb{Q}$ and with a root α. Let $\alpha_1, \dots, \alpha_p$ be all the roots of R. Then all elementary symmetric polynomials of $\alpha_1, \dots, \alpha_p$ can be expressed in terms of the coefficients of R and, therefore, are rational. Let us similarly define $\beta_1, \dots, \beta_q; \dots; \omega_1, \dots, \omega_r$ and consider the polynomial

$$P(x) = \prod_{i,j,\dots,k} (\alpha_i + \beta_j x + \dots + \omega_k x^n).$$

This polynomial is nonzero and all its coefficients can be expressed in terms of the elementary symmetric polynomials $\sigma_s(\alpha_1, \dots, \alpha_p)$, ..., $\sigma_t(\omega_1, \dots, \omega_r)$; hence, its coefficients are rational. Since $P(x_0) = 0$, it follows that $x_0 \in \overline{\mathbb{Q}}$.

In the general case the following statement holds for the automorphisms of an algebraically closed field Ω over its subfield K.

4.2.4. THEOREM. *If the elements* $x, y \in \Omega$ *are transcendental over* K, *then there exists an automorphism of* Ω *over* K *that sends* x *into* y. *If the elements*

$x, y \in \Omega$ are roots of the same irreducible polynomial over K, then there exists an automorphism of Ω over K that sends x into y.

We will not prove this theorem for arbitrary fields (for the proof see [**B3**], Ch. IV, §6, Prop. 3), but for the most interesting case — the automorphisms of $\mathbb{C}$ over $\mathbb{Q}$ — we will prove not only this theorem, but several of its generalizations. For example, we will prove that the cardinality of the set of automorphisms of $\mathbb{C}$ coincides with the cardinality of the set of all maps $\mathbb{C} \longrightarrow \mathbb{C}$ (i.e., it is greater than the cardinality of $\mathbb{C}$). Our exposition follows [**C15, C18**]).

First, observe that a field isomorphism $\varphi : F \longrightarrow G$ can be extended to an isomorphism $\varphi' : F(\alpha) \longrightarrow G(\beta)$ if and only if the following conditions hold:

1) if an element α is algebraic over F and P is an irreducible polynomial over F with root α, then β is a root of the polynomial $\varphi(P)$;

2) if α is transcendental over F, then β is transcendental over G.

For our arguments we will need Zorn's lemma. The point is that proofs by induction are only applicable to countable sets whereas the dimension of $\mathbb{C}$ over $\mathbb{Q}$ is uncountable. Therefore, to work with automorphisms of $\mathbb{C}$ over $\mathbb{Q}$ we need another technique and Zorn's lemma is sufficiently convenient for this purpose.

Before we formulate Zorn's lemma, let us give several definitions. Let g be a set. Denote by 2^g the set of all the subsets of g. A set $A \subset 2^g$ is called a *chain* if for any pair of its elements $a, b \in A$ either $a \subset b$ or $b \subset a$ (recall that a and b are subsets of the same set g). A set $B \subset 2^g$ is called *Zorn closed* if for any chain $A \subset B$ the set B contains also the union of all the elements of A. An element $m \in B$ is called *maximal* if the set $m \subset g$ is not contained in any other subset $a \subset g$ which is an element of B (i.e., $a \in B$).

4.2.5. ZORN'S LEMMA. *Every nonempty Zorn closed set $B \subset 2^g$ contains at least one maximal element m.*

With the help of Zorn's lemma we can extend any automorphism φ of a subfield of $\mathbb{C}$ to an automorphism of the whole field $\mathbb{C}$. For that, we have to apply Zorn's lemma to the family of automorphisms that extend φ. But there is one difficulty. It might happen that an extension of an automorphism of the field F to the field F' containing F does not leave F' invariant, so that an extension of an automorphism is not an automorphism of F' but an isomorphism of F' with some other field. For example, consider the automorphism of the field $\mathbb{Q}(\sqrt{2})$ given by the formula $a + b\sqrt{2} \mapsto a - b\sqrt{2}$. Its extension to $\mathbb{Q}(\sqrt[4]{2})$ is as follows:

$$a + b\sqrt[4]{2} + c\sqrt{2} + d\sqrt[4]{8} \;\mapsto\; a + ib\sqrt[4]{2} - c\sqrt{2} - id\sqrt[4]{8}.$$

This map is an isomorphism of $\mathbb{Q}(\sqrt[4]{2})$ to $\mathbb{Q}(i\sqrt[4]{2})$ but not an automorphism of $\mathbb{Q}(\sqrt[4]{2})$.

To overcome this difficulty let us first prove the following statement.

4.2.6. THEOREM. *Any field isomorphism $\varphi : F \longrightarrow G$ can be extended to an isomorphism $\bar{F} \longrightarrow \bar{G}$ of algebraic closures.*

PROOF. Consider all possible extensions of the isomorphism φ to an isomorphism $\varphi_\alpha : F_\alpha \longrightarrow G_\alpha$, where $F_\alpha \subset \bar{F}$ and, therefore, $G_\alpha \subset \bar{G}$. Let us show that the set

$$S = \{\text{the subsets of } \bar{F} \times \bar{G} \text{ of the form } \{(a, \varphi_\alpha(a)) \mid a \in F_\alpha\}\}$$

is Zorn closed. Consider an arbitrary chain in S. By the definition of a chain, for any two of its elements the corresponding isomorphisms φ_α and φ_β are such that one of these isomorphisms is an extension of the other. This means that to the union of all the elements of the chain an isomorphism corresponds, i.e., their union belongs to the set considered.

By Zorn's lemma the set S has a maximal element. To this element there corresponds an isomorphism $\psi : F' \longrightarrow G'$, so we must prove that $F' = \bar{F}$ and $G' = \bar{G}$. Suppose that an element $a \in F$ does not belong to F'. But a is algebraic over F' and $\bar{G}$ is algebraically closed. Therefore, $\bar{G}$ contains an element b which is the root of the image (under ψ) of the minimal polynomial of a over F'. Hence, it is possible to extend this isomorphism $\psi : F' \longrightarrow G'$ to an isomorphism $F'(a) \longrightarrow G'(b)$, but this contradicts the maximality of the element corresponding to ψ.

Thus, $F' = \bar{F}$. It remains to prove that $G' = \bar{G}$. The field G' is isomorphic to $\bar{F}$; hence, G' itself is algebraically closed. In addition, G' contains G. Hence, $G' = \bar{G}$. □

Now we can prove the theorem on extension of automorphisms of subfields of $\mathbb{C}$.

4.2.7. THEOREM. *Any automorphism φ of a subfield in $\mathbb{C}$ can be extended to an automorphism of the $\mathbb{C}$.*

PROOF. Consider all possible extensions of the given automorphism $\varphi : F \longrightarrow F$ to automorphisms $\varphi_\alpha : F_\alpha \longrightarrow F_\alpha$, where $F_\alpha \subset \mathbb{C}$. As in the proof of Theorem 4.2.6, we see that the set consisting of sets of the form $\{(a, \varphi_\alpha(a)) \mid a \in F_\alpha\}$ has a maximal element. To this element there corresponds an automorphism $\varphi' : F' \longrightarrow F'$. We must prove that $F' = \mathbb{C}$.

Suppose a complex number a does not belong to F'. If a is algebraic over F', then by Theorem 4.2.6 there exists an extension of φ' to an automorphism of the algebraic closure of F' and the latter is strictly bigger than F'. If a is transcendental over F', then there exists an extension of φ' to an isomorphism $F'(a) \longrightarrow F'(a)$ sending a to a. In both cases we get a contradiction with the maximality of F'. Hence, $F' = \mathbb{C}$. □

Observe that it is not always possible to extend an isomorphism of two subfields of $\mathbb{C}$ to an automorphism of $\mathbb{C}$. For example, there exists an isomorphism $\mathbb{C} \longrightarrow F \subset \mathbb{C}$, where $F \neq \mathbb{C}$. Such an isomorphism is constructed as follows. Let $a_1, a_2, \dots$ be a countable set of complex numbers algebraically independent over $\mathbb{Q}$. The map $a_i \mapsto a_{i+1}$ determines an isomorphism

$$\mathbb{Q}(a_1, a_2, \dots) \longrightarrow \mathbb{Q}(a_2, a_3, \dots) \subset \mathbb{Q}(a_1, a_2, \dots).$$

Consider all possible extensions of this isomorphism to isomorphisms $\theta_\alpha : F_\alpha \longrightarrow G_\alpha$ ($F_\alpha, G_\alpha \subset \mathbb{C}$) such that a_1 is transcendental over G_α.

By Zorn's lemma the set $\{F_\alpha\}$ has a maximal element which, as is easy to show, coincides with $\mathbb{C}$. Therefore, we get an isomorphism $\mathbb{C} \longrightarrow F \subset \mathbb{C}$, where the field F does not contain a_1 and, therefore, $F \neq \mathbb{C}$.

The proof of Theorem 4.2.4 for the case of automorphisms of $\mathbb{C}$ over $\mathbb{Q}$ is not a problem now. Indeed, if the complex numbers x and y are transcendental, we can consider an automorphism of the field $\mathbb{Q}(x, y)$ that interchanges x with y. By

Theorem 4.2.7 it is possible to extend this automorphism to an automorphism of the field $\mathbb{C}$. If x and y are roots of the same irreducible polynomial over $\mathbb{Q}$, then there exists a field isomorphism $\mathbb{Q}(x) \longrightarrow \mathbb{Q}(y)$ sending x to y. By Theorem 4.2.6 this isomorphism can be extended to an isomorphism of the algebraic closures of $\mathbb{Q}(x)$ and $\mathbb{Q}(y)$. But the algebraic closures of these fields coincide and, therefore, we get not just an isomorphism, but an automorphism. This automorphism can be extended to an automorphism of $\mathbb{C}$.

It is crystal clear now that the cardinality of the set of automorphisms of $\mathbb{C}$ is not less than the cardinality of the continuum. It turns out that the cardinality of the set of all automorphisms of $\mathbb{C}$ is, actually, greater than the cardinality of the continuum.

4.2.8. THEOREM. *The cardinality of the set of all automorphisms of* $\mathbb{C}$ *coincides with the cardinality of all maps* $\mathbb{C} \longrightarrow \mathbb{C}$.

PROOF. We have to prove that the cardinality of the set of all automorphisms of $\mathbb{C}$ coincides with the cardinality of the continuum. It suffices to prove that the cardinality of the set of all the automorphisms of $\mathbb{C}$ is not less than the cardinality of the set of all the subsets of the continuum. A set $B \subset \mathbb{C}$ is called a *basis of transcendentality* over $\mathbb{Q}$ if B is algebraically independent over $\mathbb{Q}$ and B is not contained in any other set of complex numbers algebraically independent over $\mathbb{Q}$. The maximality of B implies that $\mathbb{C}$ is algebraic over $\mathbb{Q}(B)$. Therefore, in particular, the cardinality of B is equal to that of the continuum.

Let us show that to any subset $S \subset B$ we can assign an automorphism φ_S of $\mathbb{C}$ such that to different sets there correspond different automorphisms.

Consider an automorphism of $\mathbb{Q}(B)$ over $\mathbb{Q}$ that sends $x \in B$ to x if $x \in S$ and to $-x$ if $x \notin S$. By Theorem 4.2.7 this automorphism can be extended to an automorphism φ_S of the whole $\mathbb{C}$. Clearly, if x belongs to one of the sets S or T and does not belong to the other set, then $\varphi_S(x) = -\varphi_T(x)$ and, therefore, $\varphi_S \neq \varphi_T$.$\square$

PROBLEMS

4.2.1. Prove that any automorphism of the field $\mathbb{R}$ is the identity one.

HINT. Any automorphism σ of $\mathbb{R}$ should preserve the elements of $\mathbb{Q}$. Moreover, the inequality $x \geq y$ is equivalent to the fact that $x - y = a^2$ for $a \in \mathbb{R}$.

4.2.2. Let f be an irreducible polynomial over $\mathbb{Q}$ and $\cos(2k\pi/n)$ be a root. Prove that all the roots of f are real ones.

HINT. Any automorphism of $\overline{\mathbb{Q}}$ over $\mathbb{Q}$ preserves the field $K_n \subset \mathbb{R}$. (For the definition of K_n see the beginning of §4.2.)

§4.3. The equation for the division of the lemniscate

Let us compute the arc length of the lemniscate $r^2 = \cos 2\theta$. If $x = r\cos\theta$ and $y = r\sin\theta$, then

$$dx^2 + dy^2 = (\cos\theta dr - \sin\theta r d\theta)^2 + (\sin\theta dr + \cos\theta r d\theta)^2 = dr^2 + r^2 d\theta^2.$$

Therefore, the differential ds of the arc length can be computed in polar coordinates by the formula $ds^2 = dr^2 + r^2 d\theta^2$.

In our case $2r\,dr = -2\sin\theta\,d\theta$. Therefore,

$$dr^2 + r^2 d\theta^2 = dr^2 + \frac{r^4 dr^2}{1-\cos^2 2\theta} = \frac{dr^2}{1-r^4}.$$

Hence,

$$s(r) = \int_0^r \frac{dx}{\sqrt{1-x^4}} = \int_0^r \frac{dx}{\sqrt{(1-x^2)(1-k^2x^2)}}, \quad \text{where } k^2 = -1.$$

Thus, $r = \operatorname{sn} s$ is the elliptic Jacobi sine function for $k^2 = -1$. It is true that one does not usually consider the Jacobi sine with such a modulus k; still, the addition theorem is, actually, true in this case as well:

$$\operatorname{sn}(u+v) = \frac{\operatorname{sn}\ u \operatorname{cn}\ v \operatorname{dn}\ v + \operatorname{sn}\ v \operatorname{cn}\ u \operatorname{dn}\ u}{1 + \operatorname{sn}^2 u \operatorname{sn}^2 v},$$

where $\operatorname{cn}\ u = \sqrt{1-\operatorname{sn}^2 u}$ and $\operatorname{dn}\ u = \sqrt{1+\operatorname{sn}^2 u}$.

In what follows we set

$$\varphi(s) = \operatorname{sn}\ s \quad \text{for } k^2 = -1\ .$$

The relation

$$ds = \frac{dr}{\sqrt{1-r^4}} = \frac{d\varphi(s)}{\sqrt{1-\varphi^4(s)}}$$

means that $\varphi' = \sqrt{1-\varphi^4}$. Therefore, the addition theorem can be expressed in the form

$$\varphi(u+v) = \frac{\varphi(u)\varphi'(v) + \varphi(v)\varphi'(u)}{1+\varphi^2(u)\varphi^2(v)}.$$

The function φ possesses a very important property which we will repeatedly use:

$$\varphi(iu) = i\varphi(u).$$

Indeed, setting $x = iy$ we get

$$\int_0^{ir} \frac{dx}{\sqrt{1-x^4}} = i\int_0^r \frac{dy}{\sqrt{1-y^4}}.$$

Hence, if

$$u = \int_0^r \frac{dy}{\sqrt{1-y^4}},$$

then $r = \varphi(u)$ and $ir = \varphi(iu)$. The relation $\varphi(iu) = i\varphi(u)$ implies that $\varphi'(iu) = \varphi(u)$ and $\varphi'(-u) = \varphi(u)$.

The relation $\varphi(iu) = i\varphi(u)$ means that the lattice of periods of the function φ turns into itself under the map $u \longmapsto iu$. Let us find out what the lattice of periods looks like. Let

$$\frac{\omega}{2} = \int_0^1 \frac{dx}{\sqrt{1-x^4}}.$$

Then the length of the lemniscate is equal to 2ω. The number ω plays the same role for the lemniscate as the number π plays for the circle.

By definition, $\varphi(\omega/2) = 1$ and $\varphi'(\omega/2) = 0$. The properties of the function sn imply that $\varphi(u+\omega) = -\varphi(u)$. Hence, $\varphi(u+i\omega) = -\varphi(u)$ and, therefore, $\varphi(u+\omega \pm i\omega) = \varphi(u)$, i.e., $\omega(1\pm i)$ are periods of φ.

Let us now find the zeros and poles of φ. Let $\alpha, \beta \in \mathbb{R}$. Then

$$\varphi(\alpha + i\beta) = \frac{\varphi(\alpha)\varphi'(\beta) + i\varphi'(\alpha)\beta(\beta)}{1 - \varphi^2(\alpha)\varphi^2(\beta)},$$

where all the quantities in the numerator and the denominator are finite. Hence, the equation $\varphi(\alpha+i\beta) = 0$ holds only when $\varphi(\alpha)\varphi'(\beta) = \varphi'(\alpha)\beta(\beta) = 0$. The real zeros of the functions φ and φ' are, respectively, of the form $m\omega$ and $(m+\frac{1}{2})\omega$, where $m \in \mathbb{Z}$. Hence, the zeros of φ should be of the form $m\omega + ni\omega$ or $(m+\frac{1}{2})\omega + (n+\frac{1}{2})i\omega$. Clearly, $\varphi(m\omega + ni\omega) = 0$. The equation

$$\varphi\left(u + \frac{\omega}{2}\right)\varphi\left(u + \frac{i\omega}{2}\right) = \frac{\varphi'(u)}{1+\varphi^2(u)} \cdot \frac{i\varphi'(u)}{1-\varphi^2(u)} = i$$

shows that if $\varphi(u+\frac{\omega}{2}) = 0$, then $\varphi(u+\frac{i\omega}{2}) = \infty$. Hence, $(m+\frac{1}{2})\omega + (n+\frac{1}{2})i\omega$ are the poles of the function φ. The system of zeros and poles of φ is plotted on Figure 32 (the poles are denoted by crosses); the parallelogram formed by the periods is shaded.

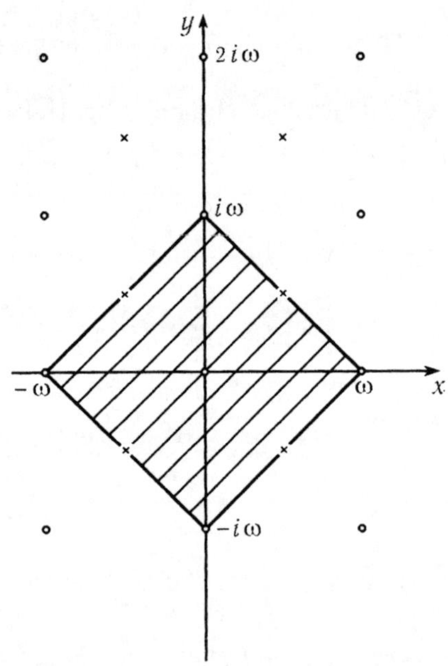

FIGURE 32

Let us return to the problem of dividing the lemniscate into equal parts. Let the length of the arc of the lemniscate between the origin and the point (r, θ) be equal to s. Then $r = \varphi(s)$. The length of one petal of the lemniscate is equal to ω and, therefore, the points for which $s = k\omega/n$, where $0 \le k < n$, divide it into n equal parts (Figure 33).

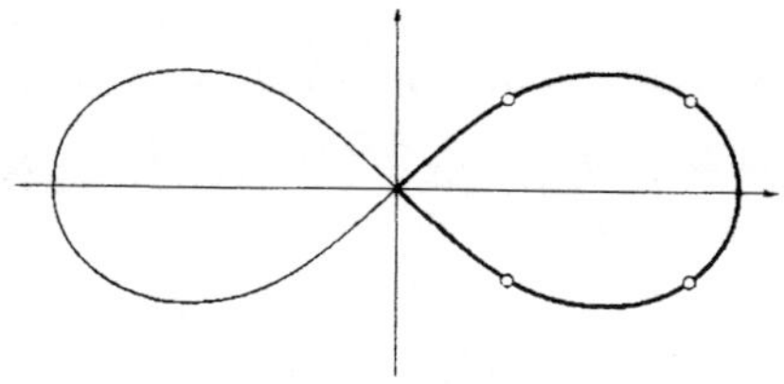

FIGURE 33

To construct these points, it suffices to construct segments of length $r = \varphi(\alpha)$, where $\alpha = k\omega/n$. Indeed, since $\cos 2\theta = r^2$, it is possible to construct the point (r, θ) with a ruler and compass if the segment r is known. Since $n\alpha = k\omega$, it follows that $\varphi(n\alpha) = 0$. The addition theorem for the function φ enables us to express $\varphi(n\alpha)$ in terms of $\varphi(\alpha)$, i.e., $\varphi(n\alpha) = F_n(\varphi)$, where $\varphi = \varphi(\alpha)$ and F_n is an algebraic function. The problem of the division of the lemniscate reduces then to that of solving the equation $F_n(\varphi) = 0$, and the division of the arc of the lemniscate between the origin and the point (r, θ) into n equal parts reduces to the problem of solving the equation $F_n(\varphi) = r$.

Let us show that the equation $F_2(\varphi) = r$ can be solved in square roots, i.e., that the arc of the lemniscate can be divided in half with a ruler and compass. By the addition theorem

$$F_2(\varphi) = \varphi(2\alpha) = \frac{2\varphi(\alpha)\varphi'(\alpha)}{1 + \varphi^4(\alpha)} = \frac{2\varphi\sqrt{1-\varphi^4}}{1+\varphi^4}.$$

Squaring the equation $2\varphi\sqrt{1-\varphi^4} = r(1+\varphi^4)$ and setting $x = \varphi^2 - \varphi^{-2}$ we get

$$x^2 + 4r^{-2}x + 4 = 0.$$

We define *odd complex numbers* to be numbers of the form $a + ib$, where a and b are integers such that $a + b$ is odd. The function $F_n(\varphi)$ is, generally, algebraic. But if n is an odd complex number, then this function is actually rational. To verify this, observe first that

$$\psi = \varphi(\alpha \pm i\alpha) = (1 \pm i)\frac{\varphi}{\sqrt{1-\varphi^4}}$$

and

$$\varphi'(\pm\alpha \pm i\alpha) = \sqrt{1-\psi^4} = \frac{1+\varphi^4}{1-\varphi^4}.$$

(In the latter formula the sign of the root is determined by the condition $\varphi'(0) = 1$.) Moreover, adding together the expressions for $\varphi(x+y)$ and $\varphi(x-y)$ we get

$$\varphi(x+y) + \varphi(x-y) = \frac{2\varphi(x)\varphi'(y)}{1 + \varphi^2(x)\varphi^2(y)}.$$

Let $y = \pm\alpha \pm i\alpha$. Then

$$\varphi(x+y) + \varphi(x-y) = \frac{2\varphi(x)(1+\varphi^4)}{1 - \varphi^4 \pm 2i\varphi^2(x)\varphi^2}.$$

Therefore, if $\varphi(x)$ and $\varphi(x-y)$ are rational functions of φ, then $\varphi(x+y)$ is also a rational function of $\varphi=\varphi(\alpha)$. For instance, setting $x=\alpha$ and $y=\alpha\pm i\alpha$, we get

$$\varphi(2\alpha+i\alpha)=\varphi\cdot\frac{-i\varphi^4+(2+i)}{1-(1-2i)\varphi^4},\quad \varphi(2\alpha-i\alpha)=\varphi\cdot\frac{i\varphi^4+(2-i)}{1-(1+2i)\varphi^4}.$$

Similar arguments show that if $a+b$ is an odd complex number, then

$$\varphi((a+bi)\alpha)=\varphi\cdot\frac{P(\varphi^4)}{Q(\varphi^4)},\tag{3.1}$$

where P and Q are polynomials with coefficients from the ring $\mathbb{Z}[i]=\{p+qi \mid p,q\in\mathbb{Z}\}$. If $b=0$ in (3.1), then the coefficients of P and Q are integers.

The polynomials P and Q possess certain interesting properties. But before we start discussing these properties in the general case, let us compute $\varphi(3\alpha)$ and $\varphi(5\alpha)$. Since

$$\varphi(2\alpha+\alpha)+\varphi(2\alpha-\alpha)=\frac{2\varphi(2\alpha)\varphi'(\alpha)}{1+\varphi^2(2\alpha)\varphi^2(\alpha)},$$

it follows that

$$\varphi(3\alpha)=-\varphi\cdot\frac{\varphi^8+6\varphi^4-3}{1+6\varphi^4-3\varphi^8}.$$

Therefore, the division of the lemniscate into three equal parts is reduced to solving the equation $\varphi^8+6\varphi^4-3=0$. Clearly, this equation is solvable in square roots.

To compute $\varphi(5\alpha)$, we can make use of the fact that

$$\varphi(3\alpha+2\alpha)+\varphi(3\alpha-2\alpha)=\frac{2\varphi(3\alpha)\varphi'(2\alpha)}{1+\varphi^2(3\alpha)\varphi^2(2\alpha)}.$$

Simple but cumbersome calculations show that

$$\varphi(5\alpha)=\varphi\cdot\frac{\varphi^{24}+50\varphi^{20}-125\varphi^{16}+300\varphi^{12}-105\varphi^8-62\varphi^4+5}{1+50\varphi^4-125\varphi^8+300\varphi^{12}-105\varphi^{16}-62\varphi^{20}+5\varphi^{24}}.\tag{3.2}$$

To solve the equation $F_5(\varphi)=0$ in square roots, Gauss used the fact that over $\mathbb{Z}[i]$ the number 5 factors into the product of $2+i$ and $2-i$. Let $\beta=(2+i)\alpha$ and $\overline{\beta}=(2-i)\alpha$. Then

$$\varphi(\beta)=\varphi\cdot\frac{-i\varphi^4+(2+i)}{1-(1-2i)\varphi^4}=\psi,\qquad \varphi(\overline{\beta})=\varphi\cdot\frac{i\varphi^4+(2-i)}{1-(1+2i)\varphi^4}=\overline{\psi},$$

$$\varphi(5\alpha)=\psi\cdot\frac{i\psi^4+(2-i)}{1-(1+2i)\psi^4},\qquad \varphi(5\alpha)=\overline{\psi}\cdot\frac{-i\overline{\psi}^4+(2+i)}{1-(1-2i)\overline{\psi}^4}.$$

Observe that the numerator in (3.2) is divisible by the numerators of the fractions ψ and $\overline{\psi}$. Dividing the numerator of (3.2) by

$$(-i\varphi^4+(2+i))(i\varphi^4+(2-i))=\varphi^8-2\varphi^4+5,$$

we get the polynomial

$$\varphi^{16}+52\varphi^{12}-26\varphi^8-12\varphi^4+1.$$

The solution of $\varphi(5\alpha) = 0$ (if we disregard the obvious case $\psi = 0$) can be obtained by, first, solving the equation $i\psi^4 + (2 - i) = 0$ and, next, solving the equation

$$\varphi\frac{-i\varphi^4 + (2+i)}{1 - (1-2i)\varphi^4} = \psi = \sqrt[4]{1+2i}. \tag{3.3}$$

One can, alternatively, solve the equation $-i\overline{\psi}^4 + (2+i) = 0$ and then the equation

$$\varphi\frac{i\varphi^4 + (2-i)}{1 - (1+2i)\varphi^4} = \overline{\psi} = \sqrt[4]{1-2i}. \tag{3.4}$$

Dividing (3.3) by (3.4) we get a quadratic equation for φ^4.

Now we prove certain properties of the polynomials P and Q for arbitrary odd complex numbers $m = a + bi$. Any such number m is, up to a summand of the form $2(\pm 1 \pm i)$, equal to either ± 1 or $\pm i$, i.e., is equal to i^ε.

4.3.1. THEOREM. $\varphi(m\alpha) = i^\varepsilon\varphi\frac{\varphi^{p-1} + A_1\varphi^{p-5} + \cdots + A_{(p-1)/4}}{1 + A_1\varphi^4 + \cdots + A_{(p-1)/4}\varphi^{p-1}}$, *where* $\varphi = \varphi(\alpha)$ *and* $p = a^2 + b^2$ *is the square of the norm of* m.

PROOF. Let us make use of the relation

$$\varphi\left(u + \frac{i^\varepsilon\omega}{2}\right)\varphi\left(u + \frac{i^{\varepsilon+1}\omega}{2}\right) = (-1)^\varepsilon i.$$

Set $x = \varphi(u + \frac{\omega}{2})$ and $y = \varphi(u + \frac{i\omega}{2})$. Then $xy = i$ and

$$\varphi\left(mu + \frac{i^\varepsilon\omega}{2}\right)\varphi\left(mu + \frac{i^{\varepsilon+1}\omega}{2}\right) = (-1)^\varepsilon i,$$

i.e.,

$$x\frac{P(x)}{Q(x)}y\frac{P(y)}{Q(y)} = (-1)^\varepsilon i.$$

Hence, $\frac{P(x)}{Q(x)} \cdot \frac{P(ix^{-1})}{Q(ix^{-1})} = (-1)^\varepsilon$. Moreover, $P(ix^{-1}) = P(x^{-1})$ and $Q(ix^{-1}) = Q(x^{-1})$, since P and Q only depend on x^4. Therefore, both P and Q are of the same degree r and $Q(x) = \lambda x^r P(x^{-1})$. We can find the coefficient λ using the fact that $\varphi(\frac{\omega}{2}) = 1$ and $\varphi(\frac{m\omega}{2}) = i^\varepsilon$. Indeed, this means that $i^\varepsilon = \frac{P(1)}{\lambda P(1)}$, i.e., $\lambda = i^{-\varepsilon}$.

It remains to demonstrate that $r = a^2 + b^2 - 1$. Under the transformation $z \longmapsto \frac{z}{m}$ the lattice generated by ω and $i\omega$ passes into the lattice of zeros of the function $\varphi(m\alpha)$. Therefore, the zeros of $\varphi(m\alpha)$ form a lattice with the area of the fundamental parallelogram equal to $\omega^2|m|^{-2}$. The area of the parallelogram of periods of the function $\varphi(\alpha)$ is equal to $2\omega^2$; hence, it contains $2\omega^2 : \omega^2|m|^{-2} = 2|m|^2$ zeros of the function $\varphi(m\alpha) = \varphi\frac{P(\varphi)}{Q(\varphi)}$. Since the degrees of P and Q are equal, it follows that $\varphi(m\alpha) = 0$ holds if and only if either $\varphi = 0$ or $P(\varphi) = 0$. Each equation $\varphi(\alpha) = a$ for various a has precisely two solutions inside the parallelogram of periods and, therefore, the degree of P is equal to $\frac{2|m|^2-2}{2} = |m|^2 - 1$. □

The roots of the equation $P(\varphi) = 0$ for odd complex m can be described as follows. Let $\Omega = \omega(1 - i)$ and $\Omega' = \omega(1 + i)$ be the periods of $\varphi(\alpha)$. The equation

$\varphi(m\alpha) = 0$ implies that

$$m\alpha = (a+bi)\omega = p\Omega + p'\Omega',$$

where $p = \frac{a-b}{2}$, $p' = \frac{a+b}{2}$. Since $\Omega' = i\Omega$, it follows that $m\alpha = (p+ip')\Omega$. Denoting the complex number $p+ip' \in \mathbb{Z}[i]$ by ν we get $m\alpha = \nu\Omega$, i.e., $\alpha = \frac{\nu\Omega}{m}$. Let ν run over the complete system of residues modulo m except the origin. Then the numbers $\varphi(\frac{\nu\Omega}{m})$ are the roots of equation $P(\varphi) = 0$. All these roots are distinct. Indeed, the equality

$$\varphi\left(\frac{\nu\Omega}{m}\right) = \varphi\left(\frac{\nu'\Omega}{m}\right)$$

is possible in two cases only: either when $\nu \equiv \nu' \pmod{m}$ (which is excluded) or when $2(\nu+\nu') = (1+i)(1-2i\nu)m$, which is also impossible because the right-hand side is only divisible by $1+i$ whereas the left-hand side is only divisible by 2. Thus, all the roots of equation $P(\varphi) = 0$ are of the form

$$x_\nu = \varphi\left(\frac{\nu\Omega}{m}\right),$$

where ν runs over the complete system of residues modulo m without the zero residue.

The example of calculations of $\varphi(m\alpha)$ for $m = 5$ already demonstrates that the arithmetic properties of m over the ring $\mathbb{Z}[i]$ are more important than those over $\mathbb{Z}$. This observation is supported by the following theorem proved by **Eisenstein**.

4.3.2. THEOREM. *Let $m = a + bi$ be an odd complex number, prime over $\mathbb{Z}[i]$. Then the numbers $A_1, \dots, A_{(p-1)/4}$ are divisible by m.*

PROOF. Dividing the polynomial $A_{(p-1)/4} + A_{(p-5)/4}\varphi^4 + \cdots + \varphi^{p-1}$ by the polynomial $1 + A_1\varphi^4 + \cdots + A_{(p-1)/4}\varphi^{p-1}$ we get

$$\varphi(m\alpha) = \varphi(c_0 + c_1\varphi^4 + c_2\varphi^8 + \cdots), \tag{3.5}$$

where $c_j \in \mathbb{Z}[i]$ and $c_0 = A_{(p-1)/4}$.

Another expression for $\varphi(m\alpha)$ can be obtained if we use that

$$\varphi'(\alpha) = \sqrt{1-\varphi^4(\alpha)} = 1 + \sum s_j\varphi^{4j}(\alpha), \tag{3.6}$$

where $s_j \in \mathbb{Q}$. Indeed, since $\varphi(0) = 0$ and $\varphi'(0) = 1$, it follows that $\varphi(\alpha) = \alpha(1+\sum p_j\alpha^j)$. The equation $\varphi'^2 = 1 - \varphi^4$ can only be fulfilled if $p_j = 0$ for $j \not\equiv 0 \pmod 4$. Hence, $\varphi(\alpha) = \alpha(1+d_1\alpha^4+d_2\alpha^8+\cdots)$, where $d_j \in \mathbb{Q}$. Replacing α with $m\alpha$, we get

$$\varphi(m\alpha) = m\alpha(1 + d_1(m\alpha)^4 + d_2(m\alpha)^8 + \cdots). \tag{3.7}$$

To compare (3.7) with (3.5), observe also that

$$\varphi^k = \varphi^k(\alpha) = \alpha^k(1 + d_1\alpha^4 + d_2\alpha^8 + \cdots)^k.$$

Therefore, (3.5) can be rewritten in the form

$$\varphi(m\alpha) = c_0\alpha(1 + d_1\alpha^4 + d_2\alpha^8 + \cdots) + c_1\alpha^5(1 + e_1\alpha^4 + e_2\alpha^8 + \cdots) + \cdots. \tag{3.8}$$

Comparing (3.7) and (3.8), we get

$$\begin{aligned} m &= c_0, \\ d_1 m^5 &= c_0 d_1 + c_1, \\ d_2 m^9 &= c_0 d_2 + c_1 e_1 + c_2, \\ d_3 m^{13} &= c_0 d_3 + c_1 e_2 + c_2 f_1 + c_3, \text{ etc.} \end{aligned}$$

It follows that $c_k = mH_k(m)l^{-1}$, where H_k is a polynomial of degree $4k$ with integer coefficients all of which are relatively prime with an integer l. The formula $nH_k(n)l^{-1} = c_k(n)$ holds for any odd complex $n \in \mathbb{Z}[i]$ (not necessarily prime); and $c_k(n) \in \mathbb{Z}[i]$. Hence, $nH_k(n) \equiv 0 \pmod{l}$.

Let $q \in \mathbb{Z}[i]$ be a prime odd complex divisor of l. Then $nH_k(n) \equiv 0 \pmod{q}$ for any odd complex number n. On the other hand, the congruence $xH_k(x) \equiv 0 \pmod{q}$ cannot have more than $\deg(xH_k(x)) = 4k+1$ distinct (modulo q) solutions.

For any odd complex q the residues after the division of odd complex numbers by q form a complete system of residues, because if n_1 is even, $n = n_1 + q$ is an odd complex number. The lattice of numbers proportional to q is generated by vectors q and iq. The area of the fundamental parallelogram of this lattice is equal to $|q|^2$; hence, the complete system of residues modulo q contains $|q|^2$ elements. Therefore, $|q|^2 \leq 4k+1$. Thus, if $4k+1 < |m|^2$, then l is not divisible by m. Since $mH_k(m)$ is divisible by l, it follows that $H_k(m)$ is divisible by l and c_k is divisible by m.

Thus, if m is an odd prime, then $c_1, c_2, \dots, c_{(p-5)/4}$, where $p = |m|^2$ are divisible by m. The relation

$$\begin{aligned} &A_{(p-1)/4} + A_{(p-5)/4}\varphi^4 + \cdots + \varphi^{p-1} \\ &\qquad = (1 + A_1\varphi^4 + \cdots + A_{(p-1)/4}\varphi^{p-1})(c_0 + c_1\varphi^4 + c_2\varphi^8 + \cdots) \end{aligned}$$

implies

$$\begin{aligned} A_{(p-1)/4} &= c_0, \\ A_{(p-5)/4} &= c_0 A_1 + c_1, \\ &\cdots\cdots\cdots\cdots\cdots\cdots \\ A_1 &= c_0 A_{(p-1)/4} + c_1 A_{(p-5)/4} + \cdots + c_{(p-5)/4}. \end{aligned}$$

Hence, the numbers $A_1, A_2, \dots, A_{(p-1)/4} = m$ are divisible by m. □

PROBLEMS

4.3.1. Prove that if $a, b, x, y, \in \mathbb{R}$ and $x + iy = \sqrt{a+ib}$, then x and y are expressible in terms of a and b using only square roots.

4.3.2. (*Eisenstein's theorem*) Assume that $f(x) = a_0x^n + a_1x^{n-1} + \cdots + a_n$, where $a_j \in \mathbb{Z}[i]$, and for a number m prime over $\mathbb{Z}[i]$ the coefficients $a_1, \dots, a_n$ are divisible by m whereas a_n is not divisible by m^2 and a_0 is not divisible by m. Then f is irreducible over $\mathbb{Z}[i]$.

4.3.3. Prove that if m is an odd complex number prime over $\mathbb{Z}[i]$, then the polynomial $\varphi^{p-1} + A_1\varphi^{p-5} + \cdots + A_{(p-1)/4}$ from Theorem 4.3.1 is irreducible.

§4.4. Proof of Abel's theorem on the division of the lemniscate

In this section we give two proofs of Abel's theorem. One of them — the classic one — belongs to Eisenstein, another — the modern one — to **Rosen** [**C14**]. The cornerstone of these two proofs, as well as of the original proof of Abel, is the fact that the lattice of periods of the function $\varphi(\alpha)$ is invariant with respect to multiplication by the complex unit i. This manifests itself most transparently in Rosen's proof. Therefore, we will start with it.

In Rosen's approach, instead of the lemniscatic function $\varphi(\alpha)$ the Weierstrass function $\wp(z)$ is used. It corresponds to the lattice $\Lambda = \{2a\omega + 2bi\omega \mid a, b \in \mathbb{Z}\}$. Observe that this lattice is contained in the lattice of periods of φ but does not coincide with it. At the end of this section we will show that for Λ we have $g_2 = -\frac{1}{4}$ and $g_3 = 0$, i.e.,

$$\wp'^2(z) = 4\wp^3(z) - \frac{1}{4}\wp(z).$$

The possibility to pass to the function $\wp(z)$ is related to the following statement.

4.4.1. Lemma. *If a segment of length $\wp(\alpha)$ can be constructed with a ruler and compass, then a segment of length $\varphi(\alpha)$ can also be constructed with a ruler and compass.*

Proof. Modulo Λ, the zeros of φ are of the form 0, ω , $i\omega$, $(1+i)\omega$ and its poles are of the form $\frac{(1+i)\omega}{2}$, $\frac{(3+i)\omega}{2}$, $\frac{(1+3i)\omega}{2}$, $\frac{(3+3i)\omega}{2}$. The function $\wp'(z)$ also has zeros ω, $i\omega$, $(1+i)\omega$, whereas 0 is a pole of $\wp'(z)$. Moreover,

$$\wp\left(\frac{1+i}{2}\omega\right) = \wp\left(\frac{3+3i}{2}\omega\right) \quad \text{and} \quad \wp\left(\frac{3+i}{2}\omega\right) = \wp\left(\frac{1+3i}{2}\omega\right).$$

Consider the function

$$g(z) = \frac{\wp'(z)}{(\wp(z) - \wp(\frac{1+i}{2}\omega))(\wp(z) - \wp(\frac{3+i}{2}\omega))}.$$

In a neighborhood of zero we have $\wp(z) = z^{-2} + \cdots$ and $\wp'(z) = -2z^{-3} + \cdots$; hence, $g(0) = 0$. It follows that g has the same zeros and poles as φ. Hence, $\varphi(z) = Cg(z)$.

Now we prove that it is possible to construct segments of length $\wp(\frac{\omega}{2})$, $\wp(\frac{1+i}{2}\omega)$ and $\wp(\frac{3+i}{2}\omega)$. First of all, observe that segments of length $\wp(\omega)$, $\wp(i\omega)$ and $\wp((1+i)\omega)$ can be constructed, since these numbers are the roots of the equation $4x^3 - \frac{1}{4}x = 0$. Moreover, it is easy to verify that

$$\wp(z \pm iz) = \mp\frac{i}{8}\frac{4\wp^2(z) - \frac{1}{4}}{\wp(z)}.$$

Let $\wp(\alpha)$ be given. Let us consider $x = \frac{\alpha}{(1+i)}$ and $y = \frac{x}{(1-i)} = \frac{\alpha}{2}$. To find $\wp(\frac{\alpha}{2})$, it suffices to solve the quadratic equations

$$\wp(\alpha) = -\frac{i}{8}\frac{4\wp^2(x) - \frac{1}{4}}{\wp(x)} \quad \text{and} \quad \wp(x) = \frac{i}{8}\frac{4\wp^2(y) - \frac{1}{4}}{\wp(y)}.$$

If it is possible to construct $\wp(\alpha)$, then it is also possible to construct $\wp'(\alpha) = \sqrt{4\wp^3(\alpha) - \frac{1}{4}}$. Hence, it is possible to construct $g(\frac{\omega}{2})$ and, since $\varphi(\frac{\omega}{2}) = 1$, the constant C can also be constructed. As a result, we see that if it is possible to construct $\wp(\alpha)$, then it is possible to construct $g(\alpha)$; thus, it is possible to construct $\varphi(\alpha)$. □

Therefore, to prove Abel's theorem, we must verify that if $n = 2^a p_1 \cdots p_m$, where the p_i are distinct Fermat primes, then the segments of length $\wp(\frac{k\omega}{n})$, where $k = 1, \ldots, n-1$, can be constructed.

The map $z \longmapsto (\wp(z), \wp'(z))$ can be considered as a homeomorphism of the torus $\mathbb{C}/\Lambda$ to the curve E defined by the equation $y^2 = 4x^3 - \frac{1}{4}x$. The addition of the points on the torus induces under this map an addition of points on E. The elements of the group E whose order divides n form a subgroup

$$E_n = \left\{ \left(\wp\left(\frac{2a\omega + 2bi\omega}{n}\right), \wp'\left(\frac{2a\omega + 2bi\omega}{n}\right)\right) \middle| 0 \le a, b < n \right\}.$$

The zero element corresponds to $a = b = 0$; this is the infinite point.

The group E_n is analogous to the group C_n for the circle (for the definition of C_n see the beginning of §4.2). By extending the analogy, let us show that if (a, b) and (c, d) are points of E, then

$$(a, b) + (c, d) = (f(a, b, c, d), g(a, b, c, d)),$$

where $\sigma f(u) = f(\sigma u)$ and $\sigma g(u) = g(\sigma u)$ for any automorphism σ of the field $\mathbb{C}$. The addition theorem

$$\wp(z_1 + z_2) = -\wp(z_1) - \wp(z_2) + \frac{1}{4}\left(\frac{\wp'(z_1) - \wp'(z_2)}{\wp(z_1) - \wp(z_2)}\right)^2 \tag{4.1}$$

shows that $f(a, b, c, d) = -a - c - \frac{1}{4}(\frac{b-d}{a-c})^2$ for $a \neq c$. By differentiating equation (4.1) and taking into account that $\wp''(z) = 6\wp^2(z) - \frac{1}{2}g_2 = 6\wp^2(z) - \frac{1}{8}$ we can represent g as a rational function of a, b, c, d with rational coefficients. In the case $z_1 \equiv z_2 \pmod \Lambda$ we can use the formula

$$\wp(2z) = -2\wp(z) + \frac{1}{4}\left(\frac{\wp''(z)}{\wp'(z)}\right)^2.$$

If $z_1 \equiv -z_2 \pmod \Lambda$, then formula (4.1) remains valid; one only has to consider the expressions in both parts as infinite ones.

With the help of the functions f and g we can get functions f_n and g_n for which $(f_n(x, y), g_n(x, y)) = n \cdot (x, y)$. The points of E_n are given in this way by the equations $f_n(x, y) = \infty$ and $g_n(x, y) = \infty$. For the finite points this means that the denominators of the fractions f_n and g_n vanish.

Now we can consider the field K_n generated over $\mathbb{Q}$ by the coordinates of the finite points of E_n. Repeating for E_n the same arguments as for C_n, we see that the group G_n of automorphisms of K_n over $\mathbb{Q}$ is isomorphic to a subgroup of the group $\mathrm{Aut}(E_n)$. Since

$$E_n \cong \frac{1}{n}\Lambda/\Lambda \cong \Lambda/n\Lambda \cong \mathbb{Z}/n\mathbb{Z} \oplus \mathbb{Z}/n\mathbb{Z},$$

it follows that $\mathrm{Aut}(E_n) \cong GL_2(\mathbb{Z}/n\mathbb{Z})$. In the case of a prime n the order of $GL_2(\mathbb{Z}/n\mathbb{Z})$ is equal to the number of bases of $(\mathbb{Z}/n\mathbb{Z})^2$, in other words, is equal to $(n^2 - 1)(n^2 - n)$. This number is divisible by $n(n+1)$ and, therefore, for $n \geq 2$ it cannot be a power of 2. Our arguments have reached a dead end!

We can only salvage it by a trick which we have repeatedly used in studying polynomials for the division of the lemniscate, namely, the invariance of the lattice of periods under multiplication by i. In our case this means that $\wp(iz)$ and $\wp'(iz)$ can be expressed in terms of $\wp(z)$ and $\wp'(z)$. Let us prove that $\wp(iz) = -\wp(z)$ and $\wp'(iz) = i\wp'(z)$. Indeed,

$$\wp(z) = z^{-2} + \sum_{\lambda\in\Lambda}{}' ((z-\lambda)^{-2} - \lambda^{-2}).$$

The invariance of Λ with respect to multiplication by i implies that $\wp(iz) = -\wp(z)$. By differentiating this equation we get $i\wp'(iz) = -\wp'(z)$. Therefore, the action of i on the torus $\mathbb{C}/\Lambda$ induces the i-action on E given by the formula $i(x, y) = (-x, iy)$. On the group $\Lambda/n\Lambda$ isomorphic to E_n the action of $k + il \in \mathbb{Z}[i]$ is given by the formula

$$(2a\omega + 2bi\omega) \pmod{n} \longmapsto (k+il)(2a\omega + 2bi\omega) \pmod{n}.$$

This action can be translated to the group E_n.

Let $F = \mathbb{Q}(i)$, F_n be the field generated by the coordinates of the points of E_n over F, and G_n be the group of automorphisms of F_n over F. If $\sigma \in G_n$, then $\sigma(i) = i$; hence, $\sigma(i(x, y)) = (-\sigma x, i\sigma y) = i\sigma(x, y)$. Furthermore, $\sigma((a, b)+(c, d)) = \sigma(a, b) + \sigma(c, d)$. It follows that G_n is a subgroup of the group of automorphisms of the $\mathbb{Z}[i]$-module $\Lambda/n\Lambda$.

The inverse to the map $a + ib \longmapsto (k + il)(a + ib)$, where a and b are taken modulo n, is the map

$$a + ib \longmapsto \frac{k - il}{k^2 + l^2}(a + ib).$$

This map is defined if and only if the number $k^2 + l^2$ is relatively prime to n. To obtain distinct maps, we have to assume that $0 \le k,\ l \le n - 1$. Thus, the order of the group of automorphisms of the $\mathbb{Z}[i]$-module $\Lambda/n\Lambda$ is equal to the number of pairs (k, l), where $0 \le k,\ l \le n-1$ and $k^2 + l^2$ is relatively prime to n. Let $\Phi(n)$ be the total number of such pairs. We break the computation of $\Phi(n)$ into several lemmas.

4.4.2. LEMMA. *If p and q are relatively prime, then $\Phi(pq) = \Phi(p)\Phi(q)$.*

PROOF. Let $0 \le a_1, b_1 \le p - 1$ and $0 \le a_2, b_2 \le q - 1$. Then the pairs $(a, b) = (a_1q + a_2p, b_1q + b_2p)$ constitute a complete system of the pairs of residues modulo pq. Moreover, $a^2 + b^2 = (a_1^2 + b_1^2)q^2 + (a_2^2 + b_2^2)p^2$. Therefore,[1]

$$(a^2 + b^2, pq) = 1 \Longleftrightarrow \{(a_1^2 + b_1^2, p) = 1 \text{ and } (a_2^2 + b_2^2, q) = 1\}. \qquad \square$$

4.4.3. LEMMA. *If p is a prime of the form $4k + 3$, then $\Phi(p) = p^2 - 1$.*

PROOF. We must prove that if $a^2 + b^2$ is divisible by p, then both numbers a^2 and b^2 are divisible by p. Suppose that $a^2 + b^2$ is divisible by p but at least one of the numbers a and b is not divisible by p. Then both numbers a and b are not divisible by p; hence, by Fermat's small theorem, $a^{p-1} \equiv 1 \pmod{p}$ and $b^{p-1} \equiv 1 \pmod{p}$, consequently, $a^{p-1} + b^{p-1} \equiv 2 \pmod{p}$. On the other hand,

[1] For brevity we denote the greatest common divisor $\mathrm{GCD}(a_1, \dots, a_n)$ of $a_1, \dots, a_n$ by $(a_1, \dots, a_n)$. In the next formula $n = 2$.

$a^{p-1} + b^{p-1} = a^{4k+2} + b^{4k+2} = (a^2)^{2k+1} + (b^2)^{2k+1}$ is divisible by $a^2 + b^2$; hence, it is divisible by p. □

4.4.4. LEMMA. *If p is a prime of the form $4k+1$, then $\Phi(p) = (p-1)^2$.*

PROOF. First of all, let us prove that any prime p of the form $4k+1$ can be represented as the sum of two squares. The simplest known proof of this statement was suggested by **D. Zagier** [**C19**].

Consider the set of all solutions of the equation $x^2 + 4yz = p$ in natural numbers. It suffices to prove that this equation has a solution for which $y = z$. In other words, the involution $\sigma(x, y, z) = (x, z, y)$ defined on the set of solutions has a fixed point. (Recall that an *involution* is a map f such that $f(f(x)) = x$ for all x; if $f(x_0) = x_0$, then x_0 is called a *fixed point*.) On a set consisting of an odd number of elements any involution has a fixed point. Therefore, it suffices to prove that the total number of solutions of the given equation is odd. To this end, it suffices to construct another involution τ of this set that has exactly one fixed point. Namely, we define:

$$\tau(x, y, z) = \begin{cases} (x + 2z, z, y - x - z) & \text{for} \quad x < y - z, & \text{(A)} \\ (2y - x, y, x - y + z) & \text{for} \quad y - z < x < 2y, & \text{(B)} \\ (x - 2y, x - y + z, y) & \text{for} \quad 2y < x. & \text{(C)} \end{cases}$$

It is easy to verify that $x \neq 2y$ and $x \neq y - z$; besides, any solution is indeed transformed by τ into a solution.

Let us divide the solutions into three types (A)–(C) according to which of the following three inequalities is satisfied:

$$x < y - z, \quad y - z < x < 2y, \quad 2y < x.$$

The map τ sends solutions of type (A) into solutions of type (C), solutions of type (B) into solutions of type (B), and solutions of type (C) into solutions of type (A). Now it is easy to verify that τ is an involution. Only a point of type (B) can be fixed. The equation $(x, y, z) = (2y - x, y, x - y + z)$ implies that $y = x$. Therefore, $p = x(x + 4z)$, i.e., $x = y = 1$ (here we have used the fact that p is prime). Thus, there is exactly one fixed point; namely, the point $(1, 1, k)$ (here we make use of the fact that p is of the form $4k + 1$).

Now we prove that for a fixed $a \neq 0$ the equation $x^2 + a^2 \equiv 0 \pmod p$ has precisely two solutions. Indeed, there exist nonzero numbers b and c such that $b^2 + c^2 \equiv 0 \pmod p$. Multiplying this inequality by $(ac^{-1})^2$ we see that $b_1^2 + a^2 \equiv 0 \pmod p$, where $b_1 = abc^{-1}$. Therefore, $x^2 \equiv b_1^2 \pmod p$; the solutions of this equation are $x = \pm b_1$. Thus, only $2(p-1)$ pairs with nonzero a and b and the pair $(0, 0)$ do not enter $\Phi(p)$. It follows that

$$\Phi(p) = p^2 - 1 - 2(p - 1) = (p - 1)^2. \qquad \square$$

It is also obvious that $\Phi(2) = 2$.

4.4.5. LEMMA. *Let p be a prime, $k \geq 1$. Then $\Phi(p^k) = (p^{k-1})^2 \Phi(p)$.*

PROOF. The numbers $a + a_1 p$, where $0 \leq a \leq p - 1$ and $0 \leq a_1 \leq p^{k-1} - 1$, form a complete system of residues modulo p^k. The number $(a + a_1 p)^2 + (b + b_1 p)^2$ is not relatively prime to p^k if and only if it is not relatively prime to p, i.e., $a^2 + b^2 \equiv 0 \pmod p$. It remains to observe that to every pair (a, b) that enters $\Phi(p)$ there corresponds $(p^{k-1})^2$ pairs that enter $\Phi(p^k)$.

Now it is easy to verify that $\Phi(n)$ is a power of 2 if and only if $n = 2^a p_1 \cdots p_m$, where the p_i are distinct Fermat primes. Indeed, $\Phi(2^a p_1^{k_1} \cdots p_m^{k_m})$ can be a power of 2 only if $k_1 = \cdots = k_m = 1$. If $p = 4k+1$, then $\Phi(p) = (p-1)^2$. This number is a power of 2 only if $p = 1 + 2^c$. Let $p = 4k+3$ and $\Phi(p) = p^2 - 1 = (p-1)(p+1)$ be a power of 2. The consecutive even numbers $p-1$ and $p+1$ can be powers of 2 only if $p = 3$. □

To complete the proof of Abel's theorem it remains to verify that for the lattice considered, i.e., $\Lambda = \{2a\omega + 2bi\omega | a, b \in \mathbb{Z}\}$, we have

$$g_2 = 60 \sum{}' (2a\omega + 2bi\omega)^{-4} = \frac{1}{4} \quad \text{and} \quad g_3 = 140 \sum{}' (2a\omega + 2bi\omega)^{-6} = 0.$$

The second equality is obvious since $g_3(\Lambda) = -g_3(i\Lambda)$ and in our case $i\Lambda = \Lambda$. The main difficulty is to prove that $\sum'(a\omega + bi\omega)^{-4} = \frac{1}{15}$.

Consider the three lattices

$$L_0 = \{a\omega + bi\omega\},$$
$$L_1 = \left\{ \frac{a\omega + bi\omega}{2} \,\middle|\, a \text{ and } b \text{ are odd} \right\},$$
$$L_2 = \left\{ \frac{a\omega + bi\omega}{2} \,\middle|\, a - b \text{ is odd} \right\}.$$

Then $\frac{1}{2}L_0 = L_0 \cup L_1 \cup L_2$ and $L_2 = \frac{1+i}{2} L_1$. Define $|L| = \sum'_{l \in L} l^{-4}$. Then

$$16|L_0| = \left|\frac{1}{2}L_0\right| = |L_0| + |L_1| + |L_2| \quad \text{and} \quad |L_2| = \left(\frac{2}{1+i}\right)^4 |L_1| = -4|L_1|;$$

hence, $|L_1| = -5|L_0|$. To prove the desired equality $|L_0| = \frac{1}{15}$ let us derive one more relation between $|L_1|$ and $|L_0|$. For this we use the fact that L_0 is the lattice of zeros of $\varphi(z)$ and L_1 is the lattice of its poles. Taking into account that $\varphi'(0) = 1$ we get

$$\varphi(z) = z \prod_{\alpha \in L_0}{}' \left(1 - \frac{z}{\alpha}\right) \prod_{\beta \in L_1}{}' \left(1 - \frac{z}{\beta}\right)^{-1},$$

where the infinite products should be understood as limits of finite products over $|\alpha|, |\beta| \leq N$ as $N \longrightarrow \infty$. The nonzero elements of the lattices L_0 and L_1 can be divided into 4-tuples of the form $\{\pm\gamma, \pm i\gamma\}$; hence,

$$\varphi(z) = z \prod{}' \left(1 - \frac{z^4}{\alpha^4}\right) \prod \left(1 - \frac{z^4}{\beta^4}\right)^{-1},$$

where $0 \leq \arg\alpha, \arg\beta < \frac{\pi}{2}$. Therefore,

$$z\frac{\varphi'(z)}{\varphi(z)} = z\frac{d}{dz}\ln\varphi(z) = 1 + (|L_1| - |L_0|)z^4 + \cdots. \tag{4.2}$$

The function $z^{-1}\varphi(z)$ does not vary under the change of z to $-z$ or to $\pm iz$; hence, $\varphi(z) = z(1 + cz^4 + \cdots)$. Moreover, $(\varphi'(z))^2 = 1 - \varphi^4(z)$. Hence,

$$(1 + 5cz^4 + \cdots)^2 = 1 - z^4(1 + cz^4 + \cdots)^4$$

and, therefore, $c = -\frac{1}{10}$. It follows that

$$(4.3) \qquad z\frac{\varphi'(z)}{\varphi(z)} = 1 + 4cz^4 + \cdots = 1 - \frac{2}{5}z^4 + \cdots .$$

Comparing (4.2) with (4.3) we deduce that $|L_1| - |L_0| = -\frac{2}{5}$. Since $|L_1| = -5|L_0|$, it follows that $|L_0| = \frac{1}{15}$. □

Now let us briefly reproduce Eisenstein's proof, more exactly, a modified version suggested by Melnikov (see [**C10**]). This proof is based on the study of the properties of the polynomials associated with the division of the lemniscate.

Thus, consider the numbers $\varphi(\frac{k\Omega}{n})$ for $k = 1, \ldots, n-1$, where n is a natural number. By the addition theorem, it suffices to confine ourselves to the case when n is a prime. For $n = 2$ the theorem was proved in §4.3, where we found an explicit form of the division equation and solved it in quadratic radicals. Therefore, suppose that n is an odd prime. If $n \equiv 3 \pmod 4$, then n remains prime in the ring $\mathbb{Z}[i]$ whereas if $n \equiv 1 \pmod 4$, then n factors in $\mathbb{Z}[i]$ in the product of two prime conjugate factors $m = a + bi$ and $m' = a - bi$.

After calculating $\varphi(\frac{k\Omega}{m})$ and $\varphi(\frac{k\Omega}{m'})$ we can find $\varphi(\frac{k\Omega}{n})$ using the addition theorem. Thus, suppose that n is a prime in $\mathbb{Z}[i]$, and consider the equation of the division of the lemniscate

$$(4.4) \qquad \Phi(x) = x^{p-1} + A_1x^{p-5} + A_2x^{p-9} + \cdots + A_{\frac{p-1}{4}} = 0.$$

By Theorem 4.3.2, for a prime Gauss number n all the coefficients A_l are divisible by n and, moreover, $A_{\frac{p-1}{4}} = n$. Applying Eisenstein's criterion we see that $\Phi(x)$ is irreducible for n prime. Let g be a primitive root modulo the prime n. Clearly, g can be selected odd.

Denote the roots of equation (4.4) by

$$x_0 = \varphi\left(\frac{\Omega}{n}\right), \quad x_1 = \varphi\left(g\frac{\Omega}{n}\right), \ldots, x_{p-2} = \varphi\left(g^{p-2}\frac{\Omega}{n}\right).$$

We see that each of the roots is the same rational function of the previous root, i.e.,

$$x_{l+1} = x_l\frac{\Phi_g(x_l)}{\Psi_g(x_l)}, \qquad l = 1, \ldots, p-3.$$

This means that the equation of the division of the lemniscate into n parts is abelian, hence solvable in radicals (for an exhaustive theory of abelian equations see the two-volume treatise by Burnside and Panton [**B5**]). All these radicals are quadratic if and only if $p = 2^\alpha + 1$. If n is a prime Gaussian number, then p is also a prime and is of the form $p = 2^{2^\beta} + 1$. If p is the square of a prime q such that $q \equiv 3 \pmod 4$, then the identity $q^2 = 2^\alpha + 1$ is only possible for $q = 3$.

§4.5. Several remarks on Serret's curves

" '*Wenn die Könige baun, haben die Kärrner zu tun*' (When kings are building, *says a German poet*, carters have work to do. — A line from xenia "Kant and his followers" ascribed to Schiller. — *But carters need roads. Not seldom, in the history of our science, has it happened that the king opened up a new road into the promised land and that his successors, intent upon their own paths, allowed it to be overrun by brambles and become unfit for transit.*"

This maxim of Andre Weil (cited from [**B25**]) is quite applicable to Serret's curves. It suffices to say that even in the famous *Lectures on the History of Mathematics in XIX Century* by Felix Klein we encounter Serret's name only once and in a quite different context.

The only mention of these curves that we managed to fish out in the mathematical literature is due to Salmon and is from Victorian times.

About two years ago Serret's curves drew the attention of the Yugoslavian mathematician A. Lipkowsky. Below we reproduce some of his results. We need certain properties of plane algebraic curves. We list these properties only briefly, referring for more details to [**B23, B4**].

Choose a projective coordinate system in the complex projective plane. Consider an irreducible homogeneous polynomial $f(x, y, z)$ of degree n in variables x, y, z with complex coefficients. If one triple of coordinates (a_1, a_2, a_3) of a point P satisfies the equation $f(x, y, z) = 0$, then any other triple of coordinates of P also satisfies this equation since $f(ra_1, ra_2, ra_3) = r^n f(a_1, a_2, a_3)$. The set of all points P with this property is called an *irreducible algebraic curve of degree n*.

Let C be a curve given by the equation $f(x, y, z) = 0$. The polynomial $f(x, y, z)$, being irreducible, is not divisible by z, so it is uniquely determined by the corresponding nonhomogeneous polynomial $f(x, y, 1)$. Denote it by $F(x, y)$. The equation $F(x, y) = 0$ is called the equation of the curve in the corresponding affine coordinate system. It is clear that the solutions of the equation $F(x, y) = 0$ correspond to those points of the curve C that do not lie on the infinite line $z = 0$.

Let $A = (a_1, a_2, a_3)$ and $B = (b_1, b_2, b_3)$ be two distinct points of the curve C. A point $P = (x_1, x_2, x_3)$ lies on the line L joining A and B if and only if $x_k = sa_k + tb_k$, $k = 1, 2, 3$, for some s and t. The values of s and t such that the corresponding point lies on the curve C are the solutions of the equation

$$f(sa + tb) \equiv f(sa_1 + tb_1, sa_2 + tb_2, sa_3 + tb_3) = 0.$$

Since f is an irreducible polynomial, $f(sa+tb)$ does not vanish identically in s and t. Therefore, it is a homogeneous polynomial of degree n in these variables, and the equation $f(sa + tb) = 0$ is satisfied for exactly n values of the ratio $s : t$ provided we take into account the multiplicities. Each such value of $s : t$ determines an intersection point of the line L and the curve C. It is convenient to regard a point corresponding to a root of multiplicity r as a multiplicity r intersection point of L and C.

Now let us analyze in more detail the intersection of the curve C with a straight line L passing through a given point P of C. Let $f(x, y) = 0$ be the equation of C in an affine coordinate system in which P has the coordinates (a, b). Then L is given by the equations

$$x = a + \lambda t,$$
$$y = b + \mu t,$$

and any line L passing through P is completely determined by the ratio $\lambda : \mu$. The intersection points of C and L correspond to the roots of the equation

$$f(a + \lambda t, b + \mu t) = 0.$$

Expanding the right-hand side of this equation in powers of t and taking into account that $f(a,b) = 0$ we obtain

$$(f_x\lambda + f_y\mu)t + (f_{xx}\lambda^2 + 2f_{xy}\lambda\mu + f_{yy}\mu^2)\frac{t^2}{2!} + \cdots = 0,$$

where f_x, f_y, etc., are the values of the partial derivatives of f at P. Two cases are possible.

(a) First, let us assume that f_x and f_y do not vanish simultaneously. Then almost all lines passing through P intersect C at P with multiplicity 1. The only exception is the line corresponding to the ratio $\lambda : \mu$ such that

$$f_x\lambda + f_y\mu = 0.$$

This line is the tangent line to C at P.

(b) Now let us assume that all partial derivatives up to order $r-1$ vanish at P and at least one derivative of order g does not vanish at P. Then each line passing through P intersects C with multiplicity at least r, and exactly r lines have the intersection of multiplicity higher than r. These exceptional curves are tangent to C at P and correspond to the solutions of the equation

$$\left(f_{x,\dots,x}\lambda^r + \binom{r}{1} f_{x,\dots,x,y}\lambda^{r-1}\mu + \cdots + f_{y,\dots,y}\mu^r \right) = 0.$$

Note that each exceptional line should be counted as many times as the multiplicity of the corresponding root. In case (b) P is called a multiplicity r point of the curve C.

A multiplicity one point of the curve C is called a *simple* point. Points of multiplicity two and more are called *singular* points. A point of multiplicity r is called an *ordinary point of multiplicity* r if at this point the curve admits r distinct tangent lines. A necessary and sufficient condition that (a,b,c) is a singular point is given by the equations

$$f(a,b,c) = \frac{\partial f(a,b,c)}{\partial x} = \frac{\partial f(a,b,c)}{\partial y} = \frac{\partial f(a,b,c)}{\partial z} = 0$$

in projective coordinates or by the equations

$$F(a,b) = \frac{\partial F(a,b)}{\partial x} = \frac{\partial F(a,b)}{\partial y} = 0$$

in affine coordinates.

The singularity criterion can be expressed in projective coordinates as follows. A point P of the curve C is a point of multiplicity r if and only if all derivatives of f of order $r-1$ vanish at P and there exists a derivative of order r that does not vanish at P.

Now let us consider two projective planes: Pl_1 with coordinates (x_1, x_2, x_3) and Pl_2 with coordinates (y_1, y_2, y_3). Define a mapping $T : Pl_1 \to Pl_2$ by the formula

$$y_i = x_j x_k,$$

where $(i,j,k) = (1,2,3)$ and all three indices i, j, k are distinct. The mapping T is called the *quadratic transformation* or *blow-up* of the plane Pl_1 to Pl_2. This mapping has the following properties.

(1) Each point of Pl_1 with the exception of the points $(1,0,0)$, $(0,1,0)$, and $(0,0,1)$ is mapped to a certain point of Pl_2. The exceptional points are called the *fundamental points* of T; the images of these points are not defined.

(2) All points of the line $x_i = 1$, $i = 1,2,3$, except the fundamental points are mapped to the point $y_i = 1$, $y_j = y_k = 0$. Three lines $x_i = 0$ are called *irregular lines* of T.

Denote by t' the mapping

$$x_i = y_j y_k$$

of Pl_2 to Pl_1. Clearly, it also possesses properties (1) and (2). Furthermore, we have the following property.

(3) If $x = (x_1, x_2, x_3)$ is a point that does not lie on an irregular line of the mapping T, then the point $y = T(x)$ does not lie on an irregular line of the mapping T', and $T'(y) = x$.

Since T' has a similar property, we can say that T and T' define a one-to-one correspondence between points of the planes Pl_1 and Pl_2 that does not lie on the corresponding irregular lines.

If $f(x_1, x_2, x_3) = 0$ is the equation of a curve on the plane T, then the images of the points of this curve under T satisfy the equation

$$g(y) = f(y_2y_3, y_1y_3, y_1y_2) = 0.$$

The curve $g = 0$ is called the *algebraic image* of the curve $f = 0$ under the mapping T. Let f be an irreducible polynomial and let g be the algebraic image of f. If $g(y) = p(y)f'(y)$, where $p(y)$ is the product of certain powers of y_i and f' is not divisible by any y_i, then the curve $f' = 0$ is called the *image* of the curve $f = 0$ under the mapping T.

The following fundamental theorem holds: *any irreducible curve can be made an irreducible curve with only ordinary singularities by applying consecutive quadratic transformations.*

To analyze singular points of a curve it is convenient to parametrize parts of this curve by formal power series. Denote by $\mathbb{C}[[t]]$ the field of formal power series in variable t over complex numbers. Each nonzero element $u \in \mathbb{C}[[t]]$ can be uniquely written in the form

$$u = (a_0 + a_1 t + \cdots)/t^k,$$

where k is an integer and $a_0 \neq 0$. The number k is called the *order of the series* u and is denoted by $O(u)$.

Let $f(x_1, x_2, x_3) = 0$ be the equation of an irreducible curve C in the complex projective plane. Elements $u_1, u_2, u_3 \in \mathbb{C}[[t]]$ define a *parameterization* of the curve C if

(1) $f(u_1, u_2, u_3) = 0$;

(2) there does not exist a nonzero element $e \in \mathbb{C}[[t]]$ such that $eu_1 \in \mathbb{C}$ for $i = 1,2,3$.

Let $u = (u_1, u_2, u_3)$ be a parameterization of the curve C and $h = -\min O(u_i)$. The elements (v_1, v_2, v_3) defined by the formula $v_i = t^h u_i$ determine the same parameterization of C. Moreover, $v_i \in \mathbb{C}[[t]]$ and at least one of the numbers $v_i(0) = a_i$ does not vanish. In this case the point $a = (a_1, a_2, a_3)$ is called the *center* of the parameterization. It is clear that under a transformation of coordinates the coordinates of the center transform in the same way as the coordinates of points. Therefore, the center of the parameterization is defined uniquely.

If $u = (u_1, u_2, u_3)$ is a parameterization and $s \in \mathbb{C}[[t]]$ is such that $O(s) > 0$, then $v = (u_1(s), u_2(s), u_3(s))$ is also a parameterization with the same center. If $O(s) = 1$, the two parameterizations are called *equivalent.*

Let us assume that $u_i \in \mathbb{C}[[t^r]]$ for some $r > 1$. Then we can use power series in the new variable $T' = t^r$. In this case the parameterization $u = (u_1, u_2, u_3)$ (and any equivalent parameterization) is called *reducible.* Otherwise the parameterization is called *irreducible.* An equivalence class of irreducible parameterizations of the curve C is called a *branch* of C. The common center of these parameterizations is called the center of the branch.

If $u = (u_1, u_2, u_3)$ is a parameterization of a branch of P such that $O(u_i) \geq 0$ for all i and $O(u_i) = 0$ for at least one i, and if $g(x_1, x_2, x_3)$ is an arbitrary irreducible polynomial, then we can define the order $O_P(g)$ of g as the order of the power series $g(u_1, u_2, u_3)$. The positive number $r = \min O_P(L)$, the minimum being taken over equations of all lines L passing through the center of the branch P, is called the *order of the branch.*

One of the most important applications of the notion of a branch and of the order is the following statement.

STATEMENT. (1) *If Q is a multiplicity r point of the curve C, then the sum of orders of all branches of C with the center Q equals r.*

(2) *A point of C is simple if and only if it is the center of just one branch.*

Now we continue the study of Serret's curves.

Recall once again the definition of Serret's curves. Let p be a fixed rational number. Consider the triangle OPM with the vertex O at the origin, the lengths of the sides OP and PM equal to $\sqrt{p}$ and $\sqrt{p+1}$, respectively, and the angles at the vertices O and M equal to α and β (see Figure 29).

Let us vary the triangle OPM so that the point O is fixed and the lengths of the sides OP and PM are fixed, whereas the angle ω between the x-axis and OM is determined from the equation

$$\cos\omega = \cos(p\alpha - (p+1)\beta). \tag{5.1}$$

Then the locus of vertices M is called *Serret's curve* corresponding to the parameter p. We denote this curve by S_p.

Let ρ be the length of the segment OM. Then the coordinates of M are $x = \rho\cos\omega$, $y = \rho\sin\omega$. The law of cosines implies that

$$\cos\alpha = \frac{\rho^2 - 1}{2\rho\sqrt{p}}, \qquad \cos\beta = \frac{\rho^2 + 1}{2\rho\sqrt{p+1}}. \tag{5.2}$$

Therefore, for p rational, the expression (5.1) can be represented as a polynomial in $\cos\alpha$, $\cos\beta$, $\sin\alpha$, and $\sin\beta$. In other words, there is a polynomial dependence between the variables x, ρ and y, ρ, i.e., a system of polynomial equations

$$P(x, \rho) = 0, \qquad Q(y, \rho) = 0. \tag{5.3}$$

Eliminating ρ from these equations we get the following polynomial equation for S_p:

$$F(x, y) = 0. \tag{5.4}$$

The following are several examples of equations (5.3) for different values of p:

$$p = 1, \qquad x = \frac{-1 + 4\rho^2 + \rho^4}{4\rho^2}, \qquad x^2 + y^2 = \rho^2; \tag{5.5}$$

$$p = 2, \qquad x = \frac{1 - 12\rho^2 + 27\rho^4 + 4\rho^6}{4 \cdot 3\sqrt{3}\rho^4}, \qquad x^2 + y^2 = \rho^2; \tag{5.6}$$

$$p = 3, \qquad x = \frac{-1 + 24\rho^2 - 162\rho^4 + 256\rho^6 + 27\rho^8}{96 \cdot \sqrt{3}\rho^6} \qquad x^2 + y^2 = \rho^2; \tag{5.7}$$

$$p = 1/2, \qquad x = \frac{-2 + 6\rho^2 + \rho^6}{3\sqrt{3}\rho^3}, \qquad x^2 + y^2 = \rho^2; \tag{5.8}$$

$$p = 1/3, \qquad x = \frac{\sqrt{3}(-9 + 24\rho^2 - 2\rho^4 + 3\rho^8)}{32\rho^4}, \qquad x^2 + y^2 = \rho^2. \tag{5.9}$$

The simplest Serret's curves are plotted in Figures 34–36.

FIGURE 34. Curve S_1

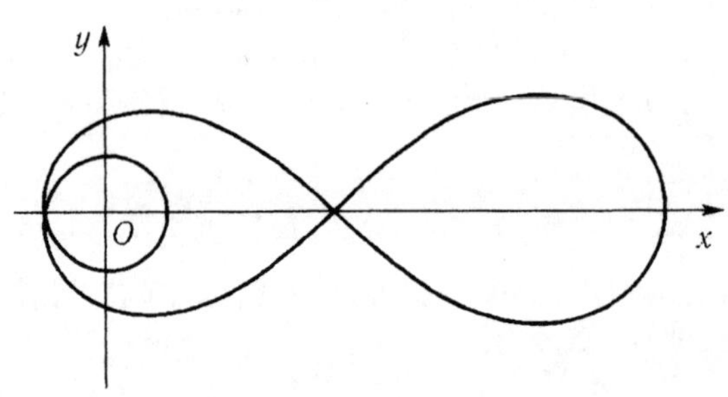

FIGURE 35. Curve S_2

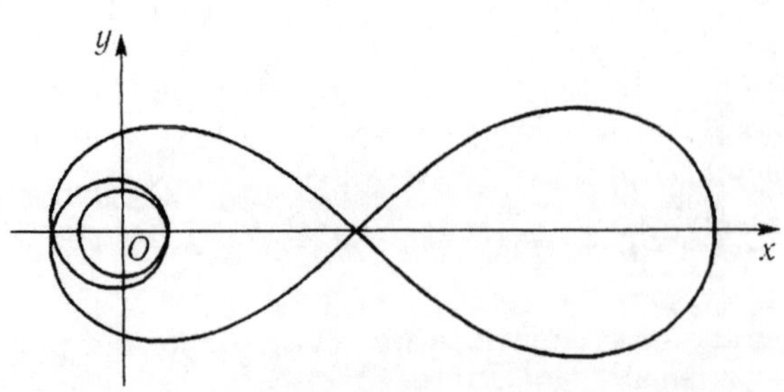

FIGURE 36. Curve S_3

Let $p = \frac{s}{t}$ $(s, t \in \mathbb{N})$ be the representation of a positive rational number p as the fraction reduced to its lowest terms. It is possible to show that relations (5.3) are of the form

$$x = \frac{a_0 + a_1\rho^2 + \cdots + a_n\rho^{2n}}{b\rho^m}, \qquad x^2 + y^2 = \rho^2, \tag{5.10}$$

where $n = s + t$ and $m = 2s + (t - 1)$. Besides, $a_0 = (-t)^s$ and $a_n = s^s$. For odd denominators t the form of the polynomial $F(x, y)$ in equation (5.4) is, therefore, as follows:

$$F(x,y) = bx(x^2+y^2)^{s+(t-1)/2} - a_0 - a_1(x^2+y^2) - \cdots - a_{s+t}(x^2+y^2)^{s+t}. \tag{5.11}$$

In this case the degree of the curve is equal to $\alpha = 2n = 2s + 2t$.

For t even the form of the polynomial $F(x, y)$ is more complicated:

$$F(x,y) = b^2x^2(x^2+y^2)^{2s+(t-1)} - [a_0 - a_1(x^2+y^2) + \cdots + a_{s+t}(x^2+y^2)^{s+t}]^2. \tag{5.12}$$

Its degree is equal to $4s + 4t$.

Here we will only consider Serret's curve S_p with integer values of p, i.e., the case of $t = 1$.

Recall (see §3.3) that for the length of an arc of Serret's curve S_p one has

$$l = \sqrt{p}\int_0^\alpha \frac{d\alpha}{\sqrt{1 - k^2\sin^2}},$$

where $k = \sqrt{\frac{p}{p+1}}$.

On the other hand, changing the notation, one has

$$dl = (\text{coefficient})\frac{dx}{y},$$

where

$$y^2 = x^4 - 2(2p+1)x^2 + 1.$$

Since the right-hand side can be factored,

$$y^2 = (x-a)(x+a)\left(x - \frac{1}{a}\right)\left(x + \frac{1}{a}\right),$$

where $a = \sqrt{p+1} + \sqrt{p}$, we can apply some birational transformation, first to lower the degree from four to three (the transformation $x - a = \frac{1}{u}$, $y = \frac{v}{u^2}$) and then to normalize the roots of the third degree polynomial on the right-hand side (a translation to have 0 as one root and a rescaling to make 1 another root). All transformations are combined together in the explicit formula

$$x \mapsto a\frac{\lambda - x}{\lambda + x}, \qquad y \mapsto 2i(a^2+1)\frac{y}{(\lambda + x)^2},$$

where $\lambda = \frac{a^2+1}{a^2-1} = \sqrt{\frac{p+1}{p}} = \frac{1}{k}$, and we obtain the elliptic curve

$$y^2 = x(x-1)(x-\lambda^2).$$

This curve is called *the elliptic curve, associated with Serret's curve* S_p. It is clear from the previous consideration that this curve is uniquely determined up to standard automorphisms of root.

It is well known that the lemniscate is a rational curve, i.e., that the lemniscate admits a rational parameterization. It turns out that all Serret's curves S_p have this property for positive integer values of p.

4.5.1. THEOREM. *The curve S_p is rational for any positive integer p.*

PROOF. This proof is based on the following statement which, in turn, follows from Hurwitz's theorem [**B8**]:

Let the plane algebraic curve be determined by the equation $F(x,y) = 0$. Let $d = \deg F$ and let ν be the multiplicity of a singular point. Set

$$g = \frac{(d-1)(d-2)}{2} - \sum \frac{\nu(\nu-1)}{2},$$

where the summation is over all singular points, infinite ones included. The curve $F(x,y) = 0$ is rational if and only if $g = 0$.

The number g is called the *genus* of the curve.

Thus, consider the polynomial (5.11),

$$F(x,y) = bx(x^2+y^2)^p - a_0 - a_1(x^2+y^2) - \cdots - a_{p+1}(x^2+y^2)^{p+1}.$$

First of all, let us study its behavior at infinity. For this, split it into homogeneous summands

$$F(x,y) = F_0(x,y) + \cdots + F_d(x,y), \qquad d = 2p+2, \tag{5.13}$$

and pass to the homogeneous polynomial of x, y, z:

$$f(x,y,z) = z^d F_0(x,y) + \cdots + zF_{d-1}(x,y) + F_d(x,y). \tag{5.14}$$

The system of equations for the singular points of $f(x,y,z)$ is as follows:

$$\frac{\partial f}{\partial x} = \frac{\partial f}{\partial y} = \frac{\partial f}{\partial z} = f = 0. \tag{5.15}$$

Therefore, the singular points on the infinite line $z = 0$ are of the form $(x : y : 0)$, where (x, y) is a solution of the equations

$$\frac{\partial F_d(x,y)}{\partial x} = \frac{\partial F_d(x,y)}{\partial y} = F_{d-1}(x,y) = 0. \tag{5.16}$$

Here

$$F_d(x,y) = -a_{p+1}(x^2+y^2)^{p+1} \quad \text{and} \quad F_{d-1}(x,y) = bx(x^2+y^2)^p.$$

Therefore, equations (5.16) can be reduced to the system

$$x(x^2+y^2)^p = y(x^2+y^2)^p = 0,$$

which yields exactly two points: $(\pm i; 1; 0)$.

Let us study the behavior of the curve at these singular points. Consider the homogeneous polynomial (5.14) in the chart $y = 1$ and translate the origin to the point $(x, z) = (\pm i, 0)$ (for the details concerning the technique used below see, for example, [**B23**] and [**B4**]). Since in both cases the calculations are similar, let us carry them through only for $(i; 1; 0)$, i.e., for $(x, z) = (i, 0)$. In this case after the transformations indicated we get the polynomial

$$b(x+i)x^p(x+2i)^p z - a_0 z^{2p+2} - a_1 x(x+2i)z^{2p} - \cdots - a_{p+1}x^{p+1}(x+2i)^{p+1}. \tag{5.17}$$

The initial form of this polynomial at the point $(0;0)$, i.e., the sum of monomials of minimal degree, is

$$b'x^pz - a_0'z^{2p+2} - a_1'xz^{2p} - \cdots - a_p'x^pz^2 - a_{p+1}'x^{p+1}, \tag{5.18}$$

where the coefficients b', a_0', ... , a_{p+1}' are determined from the coefficients b, a_0, ... , a_{p+1}. In this way we get the singular points of multiplicity $p+1$ with characteristic monomials x^pz, z^{2p+2}, x^{p+1}. The *Newton diagram* corresponding to these singular points is of the form plotted in Figure 37.

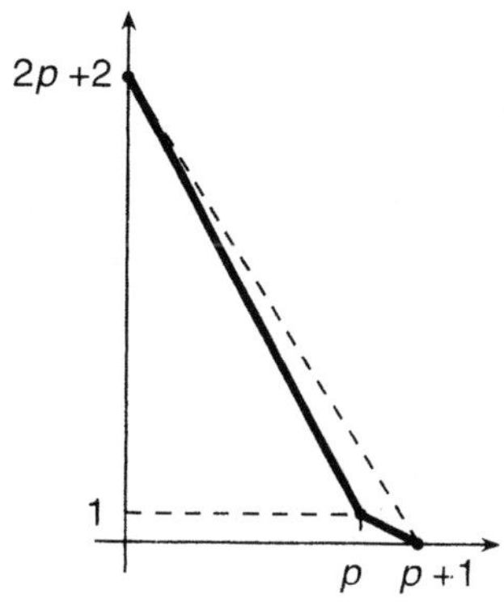

FIGURE 37

Clearly, on the sides of this diagram there are no points with integer coordinates. This means that the singularity has two branches. To compute how the genus of the singularity diminishes, we can resolve the singularity by means of blowing up. In this case just two blowups suffice. The first blowup is of the form $x \mapsto xz$, $z \mapsto z$. This implies that

$$x^pz + z^{2p+2} + x^{p+1} \mapsto z^{p+1}(x^p + z^{p+1} + x^{p+1}).$$

In the chart $x \mapsto x$, $z \mapsto zx$ there are no singularities, because the form $x^pz + z^{2p+2} + x^{p+1}$ turns into $x^{p+1}(z + x^pz^{2p} + 1)$. Thus, as a result of the first blowup we get one singular point of the form $x^p + z^{p+1} + x^{p+1}$; its multiplicity is equal to p. The second blowup yields simple points. Indeed, for $x \mapsto xz$, $z \mapsto z$ we have

$$x^p + z^{p+1} + x^{p+1} \mapsto z^p(x^p + z + x^{p+1}z).$$

If $x \mapsto x$, $z \mapsto xz$, then

$$x^p + z^{p+1} + x^{p+1} \mapsto x^p(1 + xz^{p+1} + x).$$

The tree of infinitesimally close points is of the form

$$(p+1) \to (p) \to (1),$$

and, therefore, the contribution of the two infinite points $(\pm i : 1 : 0)$ to the genus g is equal to

$$-2\left(\frac{(p+1)p}{2} + \frac{p(p-1)}{2}\right) = -2p^2.$$

Therefore, when these singularities are resolved the genus g becomes equal to

$$g = \frac{(d-1)(d-2)}{2} - \sum{}' \frac{\nu(\nu-1)}{2} - 2p^2 = p - \sum{}' \frac{\nu(\nu-1)}{2} \leq p, \tag{5.19}$$

where $\sum'$ denotes the summation over all singular points lying in the finite part of $\mathbb{C}P^2$. In particular, the inequality (5.19) implies that in the finite part the contribution of $\sum' \frac{\nu(\nu-1)}{2}$ from the singular points of S_p is not greater than p.

Now let us find the finite singular points of S_p. Making use of the symmetry of the curve, we may suppose that these singular points lie on the x-axis and, therefore, satisfy the system of equations

$$\frac{\partial F(x,0)}{\partial x} = F(x,0) = 0. \tag{5.20}$$

This assumption becomes justified if we can justify that the contribution $\sum' \frac{\nu(\nu-1)}{2}$ of these singular points is exactly equal to p.

In order to analyze the system (5.20), it is convenient to make use of the relation (5.1). The equation $F(x,0) = 0$ is equivalent to the fact that $\omega = 0$ or π which, in turn, is equivalent to the equation

$$\cos(p\alpha - (p+1)\beta) = \pm 1 \tag{5.21}$$

which implies that

$$p\alpha - (p+1)\beta = q\pi, \qquad g \in \mathbb{Z}. \tag{5.22}$$

Adding the constraints

$$\sin\beta = \sqrt{\frac{p}{p+1}} \sin\alpha \quad \text{and} \quad 0 \leq \alpha, \beta, \alpha+\beta \leq \pi \tag{5.23}$$

to (5.22) we get the system of equations:

$$\begin{cases} \beta = \frac{p}{p+1}\alpha - \frac{q}{p+1}\pi, & q \in \mathbb{Z}, \\ \beta = \arcsin(\sqrt{\frac{p}{p+1}} \sin\alpha). \end{cases} \tag{5.24}$$

For $q = 0$ the straight line $\beta = \frac{p}{p+1}\alpha$ intersects the curve $\beta = \arcsin(\sqrt{\frac{p}{p+1}} \sin\alpha)$ at exactly two points. All the other straight lines $\beta = \frac{p}{p+1}\alpha - \frac{q}{p+1}\pi$ either do not intersect the curve at all (for $q < 0$ and for $q > p$) or intersect it at exactly one point (for $q = 1, \dots, p$). The coordinates of the last intersection point are $(\pi, 0)$. Therefore, our system (5.24) or, which is the same, the initial system (5.20) has exactly 2 simple and p double roots. This means that there are no other singular points and the genus of the curve is $g = 0$. Hence, all the curves S_p for positive integer p are rational. □

We have mentioned above that the crucial moment in the proof of Abel's theorem on the lemniscate's division is the invariance of the period lattice of the lemniscatic function $\varphi(\alpha)$ with respect to the multiplication by the complex unit i. This invariance is a manifestation of an important general property of certain elliptic curves: the presence of so-called *complex multiplication*.

Let E be an elliptic curve; as a group manifold it is isomorphic to the quotient of $\mathbb{C}$ modulo the lattice Λ spanned by the periods ω_1, ω_2. The curve E is also isomorphic to the quotient of $\mathbb{C}$ modulo the lattice spanned by $z\omega_1$, $z\omega_2$ for any $z \in \mathbb{C}$ and, therefore, we may assume that $\omega_1 = 1$, $\omega_2 = \tau$ and $\operatorname{Im}\tau > 0$. Any endomorphism of E can be lifted to an automorphism of $\mathbb{C}$ and, therefore, it induces a multiplication by a complex number z such that $z\tau \in \Lambda$. The endomorphisms of E constitute the ring $A(E)$ containing the subring of the so-called *trivial automorphisms*, which is a subring isomorphic to $\mathbb{Z}$. The remaining endomorphisms, if any, are determined by complex numbers and are called *complex multiplications.* If $A(E) \neq \mathbb{Z}$, we say that E possesses complex multiplication. A generic elliptic curve E has no complex multiplications.

Indeed, suppose that z determines a nontrivial endomorphism of E. Then

$$z = a + b\tau, \quad z\tau = c + d\tau \quad (a, b, c, d \text{ are integers, } b \neq 0)$$

and, therefore,

$$a\tau + b\tau^2 = c + d\tau. \tag{5.25}$$

Hence, τ must belong to the imaginary quadratic field K and z must belong to the ring of integers $O(K)$ of K, because z determines an endomorphism of a $\mathbb{Z}$-module of finite rank. Therefore, $A(E)$ is a subring of $O(K)$ containing $\mathbb{Z}$ and has rank 2 when considered as the $\mathbb{Z}$-module. Conversely, any such subring R in the imaginary quadratic field K can be obtained in this way: it suffices to set $E = \mathbb{C}/R$.

The so-called j-invariant is the most convenient way to express that an elliptic curve E has complex multiplication. If E is determined by an equation in the Weierstrass form $y^2 = x^3 + ax + b$, then its *j-invariant* is given by the formula

$$j(E) = 1728\frac{4a^3}{4a^3 + 27b^2}. \tag{5.26}$$

Two curves are isomorphic if and only if their j-invariants are equal. If E is defined over $\mathbb{Q}$, i.e., if $a, b \in \mathbb{Q}$, there is a very simple criterion for the existence of complex multiplication: *E possesses complex multiplication if and only if $j(E)$ is an integer* ([**B12**], [**B20**]). It is possible to show ([**B20**]) that there are exactly 13 classes of elliptic curves with complex multiplication and rational j-invariant. The values of j-invariant for them are

$$\begin{gathered} 2^6 \cdot 3^3, \qquad 2^6 \cdot 5^3, \qquad 0, -3^3 \cdot 5^3, \qquad -2^{15}, \\ -2^{15}3^3, \qquad -2^{18} \cdot 3^3 \cdot 5^3, \qquad -2^{15} \cdot 3^3 \cdot 5^3 \cdot 11^3, \\ -2^{18} \cdot 3^3 \cdot 5^3 \cdot 23^3 \cdot 29^3, \qquad 2^3 \cdot 3^3 \cdot 11^3, \qquad 2^4 \cdot 3^3 \cdot 5^3, \\ 3^3 \cdot 5^3 \cdot 17^3, \qquad -3 \cdot 2^{15} \cdot 5^3. \end{gathered}$$

4.5.2. THEOREM. *Among the elliptic curves associated with Serret's curve S_p only the elliptic curve corresponding to the lemniscate S_1 possesses complex multiplication.*

PROOF. Let

$$y^2 = x(x-1)(x-\lambda^2), \quad \lambda^2 = \frac{1}{k^2} = \frac{p+1}{p},$$

be the elliptic curve associated with Serret's curve S_p. Its j-invariant is equal to

$$j = 2^8\frac{(\lambda^4 - \lambda^2 + 1)^3}{\lambda^4(\lambda^2 - 1)^2} = 2^8\frac{(p^2 + p + 1)^3}{p^2(p+1)^2}. \tag{5.27}$$

Since

$$j = 2^8 \frac{(p(p+1)+1)^3}{(p(p+1))^2} = 2^8 p(p+1) + 3 \cdot 2^8 + \frac{3 \cdot 2^8}{p(p+1)} + \frac{2^8}{p^2(p+1)^2},$$

it follows that j can only be integer for $p = 1$. In this case $j = 2^6 \cdot 3^3$. Thus, only the elliptic curve corresponding to the lemniscate S_1 admits complex multiplication. □

In conclusion let us formulate several **problems**.

First of all, it would be very useful to find rational parameterizations for Serret's curve S_p with positive integer p similar to the lemniscate's parameterization.

The main problem, however, is the study of the structure of Serret's curves for noninteger values of p. Here we have two very different families: with odd and with even values of the denominator t. It seems very plausible that for odd t all Serret's curves are rational. At the same time the computer experiments demonstrate that for even t irrational curves can occur.

Further on, it is very important to compute the j-invariants of all the corresponding elliptic curves and study whether or not they admit complex multiplication.

Finally, it would be very interesting to check whether Eisenstein's proof can be generalized to curves without complex multiplication.

CHAPTER 5

Arithmetic of Cubic Curves

In this chapter we consider certain well-known diophantine equations. Recall that a *diophantine equation* is a polynomial equation

$$f(x_1, \ldots, x_n) = 0$$

whose coefficients are integers. If this equation has a solution $\{x_1, x_2, \ldots, x_n\}$ in integers, we will say that $\{x_1, x_2, \ldots, x_n\}$ is an *integer solution.* A solution of this equation in rational numbers is called a *rational solution.* It is clear that if f is a homogeneous polynomial, then the problem of the search for rational solutions is equivalent to the problem of the search for integer solutions.

The sources of the theory of diophantine equations can be traced to the mathematics of antiquity. **Euclid** developed a method for finding the greatest common divisor d of two numbers a_1 and a_2. This method leads in general to solving the diophantine equation[1]

$$a_1x_1 + a_2x_2 = d.$$

If a number b is divisible by d, we can multiply x_1 and x_2 by b/d, getting a solution of the equation $a_1x_1 + a_2x_2 = b$. Euclid also had a method for finding the greatest common divisor (a_1, a_2, a_3) of three numbers; this method fits n numbers as well. It is based on the fact that

$$((a_1, \ldots, a_{n-1}), a_n) = (a_1, \ldots, a_n).$$

The ancient Greek mathematicians never made use of equations when they solved geometric problems. The notion of quantities as geometric objects that they developed did not allow them to do this. Consequently, the mathematicians of classical Greece were not very interested in equations. (It was **Descartes** who started to use equations systematically for the solution of geometric problems.)

After Euclid's time, no new methods for the solution of equations in integers were developed until the third century by **Diophantus**, an Alexandrian mathematician. He developed methods for solutions in integers and rational numbers of quadratic and certain cubic equations with two and more unknowns and thus constructed a foundation for a new mathematical discipline now called the *theory of diophantine equations* in his honor. He summarized his studies in a vast treatise in 13 books under the common title *Arithmetics.*

The fate of this treatise is remarkable. Soon after it was written it disappeared for more than a millennium and was considered to be lost forever. It was only in 1464 that a German scientist **Regiomontanus** accidentally found 6 of the 13 books of *Arithmetics.* It was first published in a Latin translation in 1575. After the

[1] Euclid himself did not, however, deal with solutions of such equations.

French edition, prepared by **Bachet de Mesiriac**, was published in 1621 it became a desk-top book for many mathematicians, **Pierre Fermat** and **René Descartes** among them. It was on the margins of his copy of Diophantus's *Arithmetics* that Fermat wrote one of the most notorious scholia in the history of mathematics:

> **Cubum autem in duos cubos, aut quadrato-quadratum in duos quadrato-quadratos, et generaliter nullam in infinitum ultra quadratum potestatem in duas ejusdem nominis fas est divedere; cujus rei demonstrationem mirabilem sane detexi. Hanc marginis exiguitas non caperet.**

"It is impossible to expand a cube into two cubes or a bisquare into two bisquares or, generally, any power greater than 2 into two powers with the same exponent; I have found a truly remarkable proof of this fact but the margins here are too narrow for it."

The collection of methods developed by Diophantus is still valuable. It is known under the name of the method of secants. This method allows one to completely investigate any quadratic equation and serves as a prototype for the study of cubic equations.

§5.1. Diophantus' method of secants. Second degree diophantine equations

Before we consider the general situation, let us illustrate the method of secants with a concrete example, one of those that Diophantus considers in his *Arithmetics*. Given the equation

$$x^2 - y^2 = 1, \tag{1.1}$$

we have to find all its rational solutions. Equation (1.1) singles out a hyperbola (Figure 38) in the xy-plane.

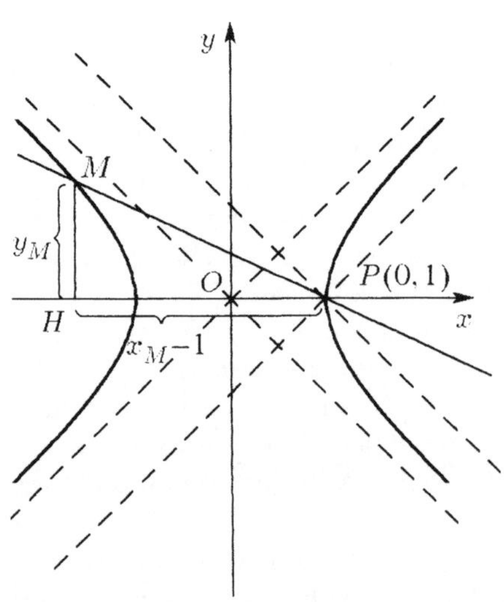

FIGURE 38

We immediately see that $(1,0)$, i.e., the intersection point P of the curve with the x-axis, is a solution. Let us draw the secant

$$y = k(x-1) \tag{1.2}$$

through this point and find the second intersection point of the secant with the curve (1.1). To do so, substitute (1.2) into (1.1) and solve the quadratic equation obtained for x. As a result we get

$$x_{1,2} = \frac{-k^2 \pm 1}{1-k^2}.$$

The root $x_1 = 1$ is known to us. It corresponds to the point $(1,\ 0)$; the second root x_2 corresponds to the second point required,

$$\left(\frac{k^2+1}{k^2-1}, \frac{2k}{k^2-1}\right). \tag{1.3}$$

For any rational $k \neq \pm 1$ this formula determines a point on our curve; hence, it determines a rational solution of the given equation; for $k = \pm 1$ the secant intersects the curve at point P only (see Figure 38). Conversely, for any rational solution, i.e., a rational point M on the curve, the secant PM is given by equation (1.2) with a rational k (because the legs of the acute triangle PMH are rational). Thus, formula (1.3) for all rational $k \neq \pm 1$ gives all solutions of equation (1.1) in rational numbers.

This method is applicable not only to the polynomial $x^2 - y^2 - 1$ but to any second degree polynomial of two variables

$$p(x,y) = Ax^2 + 2Bxy + Cy^2 + Dx + Ey + F$$

with integer or rational coefficients provided the curve $p(x,y) = 0$ has at least one rational point $P_0 = (x_0, y_0)$. Indeed, let us draw the line

$$y - y_0 = k(x - x_0) \tag{1.4}$$

through this point and find the intersection points of the curve $p(x,y) = 0$ with this line. That is, substitute (1.4) into the equation $p(x,y) = 0$. The polynomial

$$p(x, y_0 + k(x - x_0))$$

is of degree 2 with respect to x; if we set

$$p(x, y_0 + k(x-x_0)) = r(k)x^2 + s(k)x + t(k),$$

where $r(k)$, $s(k)$, $t(k)$ are polynomials of k, then for x we get an equation

$$r(k)x^2 + s(k)x + t(k) = 0.$$

We know one of the roots of this equation, namely, $x = x_0$. Therefore, the other root, $x = x_1$, can be found using the relation

$$x_0 + x_1 = -\frac{s(k)}{r(k)}.$$

Substituting the expression $x_1 = -x_0 - \frac{s(k)}{r(k)}$ in (1.4) we get

$$y_1 = y_0 - k\left(2x_0 + \frac{s(k)}{r(k)}\right).$$

Thus, the coordinates of the second point are of the form

$$(1.5)\qquad (x_1, y_1) = \left(-x_0 - \frac{s(k)}{r(k)},\quad y_0 - k\left(2x_0 + \frac{s(k)}{r(k)}\right)\right).$$

Since $s(k)$ and $r(k)$ are polynomials with rational coefficients, it follows that for any rational k such that $r(k) \neq 0$ formula (1.5) gives a rational solution of the equation $p(x, y) = 0$ and, since for rational x_1 and y_1 the slope coefficient

$$k = \frac{y_1 - y_0}{x_1 - x_0}$$

is also rational, we deduce that this formula gives all rational solutions, provided we know at least one rational solution.

The problem of existence of a rational point on a second degree curve $p(x, y) = 0$ is quite tough. It is not always the case that such points exist; for instance, there are none on the circle $x^2 + y^2 = 3$ or on the ellipse $x^2 + 82y^2 = 3$. If the curve $p(x, y) = 0$ is reducible over $\mathbb{Q}$, then the problem of finding a rational point can be reduced to the study of a linear diophantine equation and, therefore, we can confine ourselves to the study of curves $p(x, y) = 0$, where the polynomial $p(x, y) = Ax^2 + 2Bxy + Cy^2 + Dx + Ey + F$ is irreducible over $\mathbb{Q}$. It is well known that if the discriminant $\Delta = AC - B^2$ of the quadratic form $Ax^2 + 2Bxy + Cy^2$ vanishes, then the polynomial $p(x, y)$ can be reduced, by an invertible linear transformation with rational coefficients, to the form $ax^2 + y$, where $a \neq 0$, or to the form $x^2 - c$, where c is not a perfect square. If $\Delta \neq 0$, then the polynomial $p(x, y)$ can be reduced to the form

$$(1.6)\qquad ax^2 + by^2 + c,\ ab \neq 0.$$

Obviously, the curve $x^2 - c = 0$ has no rational points; it is also obvious that on the curve $ax^2 + y = 0$ there is a rational point $(0, 0)$. Thus, it remains to study the curve (1.6). If $c = 0$, then $(0, 0)$ is a rational point of the curve (1.6). Hence, we may assume that $abc \neq 0$. First of all, for equation (1.6) to be solvable in rational numbers, it is necessary that not all of the coefficients a, b, and c be of the same sign. Performing, if necessary, the change of variables

$$x \longmapsto x^{-1},\ y \longmapsto yx^{-1} \quad\text{or}\quad x \longmapsto xy^{-1},\ y \longmapsto y^{-1},$$

we can reduce the equation (1.6) to the form

$$(1.7)\qquad ax^2 + by^2 - c = 0, \quad \text{where } a > 0, b > 0, c > 0.$$

Here we can assume that a, b, and c are relatively prime and square-free integers.

If $x = p/r$, $y = q/r$, where $p, q, r \in \mathbb{Z}$, is a rational solution of equation (1.7), then the equation

$$(1.8)\qquad ax^2 + by^2 - cz^2 = 0$$

has a nonzero integer solution $\{x, y, z\} = \{p, q, r\}$. Conversely, if equation (1.8) has a nontrivial integer solution $\{x, y, z\}$, then $z \neq 0$ (otherwise we would have obtained a nontrivial solution of the equation $ax^2 + by^2 = 0$, where $a > 0$, $b > 0$) and equation (1.7) is solvable in rational numbers.

Therefore, provided that the above necessary condition is satisfied, the problem of existence of rational solutions of equation (1.6) can be reduced to the problem on nontrivial solvability in integers of equation (1.8). The numbers a, b, and c can be assumed to be not only relatively prime, but also pairwise relatively prime. Indeed, let, for example, a and b be divisible by p. Then

$$p(ax^2 + by^2 - cz^2) = a_1x_1^2 + b_1y_1^2 - c_1z^2,$$

where $x_1 = px$, $y_1 = py$, $a_1 = a/p$, $b_1 = b/p$, $c_1 = pc$.

Already **Legendre** elaborated a criterion for the existence of a nontrivial solution of equation (1.8). It goes as follows.

5.1.1. THEOREM. *If a, b and c are pairwise relatively prime and square-free natural numbers, then the equation*

$$ax^2 + by^2 - cz^2 = 0$$

has a nontrivial integer solution if and only if all three of the following congruences are solvable:

$$x^2 - bc \equiv 0 \pmod{a},$$
$$x^2 - ac \equiv 0 \pmod{b},$$
$$x^2 + ab \equiv 0 \pmod{c}.$$

It will be more convenient to pass from equation (1.8) to the equation

$$acx^2 + bcy^2 = z_1^2.$$

If (x, y, z_1) is a solution of this equation, then z_1 is divisible by c since c is square-free. Therefore, $(x, y, z_1/c)$ is a solution of equation (1.8). Thus, instead of Theorem 5.1.1 it suffices to prove the following statement.

5.1.2. THEOREM. *Let a and b be square-free natural numbers. The equation*

$$ax^2 + by^2 = z^2 \tag{1.9}$$

has a nontrivial integer solution if and only if there exist integers α, β and γ such that

(i) $\alpha^2 - a \equiv 0 \pmod{b}$,
(ii) $\beta^2 - b \equiv 0 \pmod{a}$,
(iii) $\gamma^2 + ab/h^2 \equiv 0 \pmod{h}$, *where* $h = (a, b)$.

PROOF. First, suppose that equation (1.9) has a nontrivial integer solution (x, y, z). We may assume that $\mathrm{GCD}(x, y, z) = 1$. It follows from (1.9) that $ax^2 \equiv z^2 \pmod{b}$. Let $\mathrm{GCD}(b, x) = d$. Then $x = dx_1$ and $b = db_1$; hence, $ad^2x_1^2 + db_1y^2 = z^2$. Therefore, $z = dz_1$. Thus, b_1y^2 is divisible by d, where d is square-free since b is. As a result, we see that all the numbers x, y, z are divisible by d, i.e., $d = 1$. Therefore, there exists a number x' such that $xx' \equiv 1 \pmod{b}$. Then $a \equiv \alpha^2 \pmod{b}$, where $\alpha = x'z$. The solvability of the congruence $b \equiv \beta^2 \pmod{a}$ is similarly proved.

The congruence (iii) can be rewritten in the form $\gamma^2 + a_1b_1 \equiv 0 \pmod{h}$, where $a = ha_1$, $b = hb_1$. Here the numbers a_1, b_1, h are pairwise relatively prime and square-free. Since a and b are divisible by h, it follows that $z = hz_1$; hence, $a_1x^2 + b_1y^2 = hz_1^2$. If $(x, h) = d$, then y and z_1 are divisible by d. Therefore, $d = 1$ and there exists a number x' such that $xx' \equiv 1 \pmod{h}$. By multiplying the congruence $a_1x^2 + b_1y^2 \equiv 0 \pmod{h}$ by b_1x^{12} we get $a_1b_1 + \gamma^2 \equiv 0 \pmod{h}$, where $\gamma = b_1x'y$.

Now suppose that all the three congruences (i)–(iii) are solvable. If $a = 1$, then the theorem is obvious. Moreover, we may assume that $a \geq b$ because if $b > a$

it suffices to interchange x and y. If $a = b$, then by (iii) we have $\gamma^2 + 1 \equiv 0 \pmod{a}$. Suppose that one of the prime divisors of a is of the form $p = 4k+3$. Then $\gamma^2 + 1 \equiv 0 \pmod{p}$ and

$$(\gamma^2)^{\frac{p-1}{2}} \equiv (-1)^{\frac{p-1}{2}} \equiv (-1)^{2k+1} \equiv -1 \pmod{p}.$$

On the other hand, thanks to Fermat's small theorem, $(\gamma^2)^{\frac{p-1}{2}} \equiv \gamma^{p-1} \equiv 1 \pmod{p}$. This leads to a contradiction; hence, a is the product of primes of the form $4k+1$ and, perhaps, 2. Each prime factor of a can be, therefore, represented in the form of the sum of two squares (Lemma 4.4.4). Recall the formula

$$(x^2+y^2)(z^2+t^2) = (xz-yt)^2 + (xt+yz)^2.$$

Hence, $a = r^2 + s^2$ for some integers r and s and there is a solution of equation (1.9) of the form

$$\{x, y, z\} = \{r, s, r^2 + s^2\}.$$

The rest of the proof of Theorem 5.1.2 may be sketched as follows. If $a > b > 1$, then starting with equation (1.9) we construct a new equation $Ax^2 + by^2 = z^2$, where $0 < A < a$ and for this A the congruences similar to (i)–(iii) are solvable. After a finite number of steps and interchanging A with b if $A < b$ we get either $A = b$ or $b = 1$. In each of these cases we have already shown that a solution exists. From this solution we will now construct a solution of equation (1.9).

By (ii) there exists a number β such that $\beta^2 - b = aAk^2$, where A is square-free. Here we may assume that $|\beta| \le a/2$. First, let us show that $0 < A < a$. Since $\beta^2 = aAk^2 + b < a(Ak^2+1)$, it follows that $A \ge 0$. Moreover, $b \ne 1$ and b is square-free. Hence, $\beta^2 \ne b$ and, therefore, $A \ne 0$. It is also clear that

$$A = \frac{1}{ak^2}(\beta^2 - b) < \frac{\beta^2}{ak^2} \le \frac{\beta^2}{a} \le \frac{a}{4} < a.$$

Let us now prove the solvability of the congruences required. Since $\beta^2 - b = aAk^2$, we have $\beta^2 \equiv b \pmod{A}$; hence, the congruence similar to (ii) is solvable. Moreover, $\beta = h\beta_1$, where $h = (a, b)$; hence,

$$h\beta_1^2 - b_1 = a_1Ak^2. \tag{1.10}$$

In particular, $h\beta_1^2 \equiv a_1Ak^2 \pmod{b_1}$. It follows from (i) that $a_1 \equiv h\alpha_1^2 \pmod{b_1}$. Hence, $h\beta_1^2 \equiv hA(\alpha_1 k)^2 \pmod{b_1}$. Since h, k and a_1 are relatively prime with b_1, it follows that $A \equiv p^2 \pmod{b_1}$. Moreover, it follows from (1.10) that $a_1Ak^2 \equiv -b_1 \pmod{h}$. Hence, $A(a_1k)^2 \equiv -a_1b_1 \equiv \gamma^2 \pmod{h}$. Therefore, $A \equiv q^2 \pmod{h}$. The numbers h and b_1 are relatively prime; hence, $hu + b_1v = 1$ for certain integers u and v. Consider a number $x = hup + b_1vq$. Since $hu \equiv 1 \pmod{b_1}$ and $b_1v \equiv 1 \pmod{h}$, it follows that $x \equiv p \pmod{b_1}$ and $x \equiv q \pmod{h}$; hence, $A \equiv x^2 \pmod{b_1}$ and $A \equiv x^2 \pmod{h}$. It follows that $A \equiv x^2 \pmod{b}$.

Now consider $H = \mathrm{GCD}(A, b)$. Let $A = HA_2$ and $b = Hb_2$. Then $\beta = H\beta_2$ and $H\beta_2^2 - b_2 = aA_2k^2$. Hence, $-A_2b_2 \equiv a(A_2k)^2 \pmod{H}$. It follows from (iii) that $a \equiv \alpha^2 \pmod{H}$. Hence, $-A_2b_2 \equiv r^2 \pmod{H}$.

Now suppose that the equation $AX^2 + bY^2 = Z^2$ has a nontrivial integer solution. Let us multiply both sides of the equation $AX^2 = Z^2 - bY^2$ by $aAk^2 = \beta^2 - b$ term-wise. As a result, we get

$$a(AXk)^2 = (Z^2 - bY^2)(\beta^2 - b) = (\beta Z + bY)^2 - b(\beta Y + Z)^2.$$

Hence, equation (1.9) has an integer solution $x = AXk$, $y = \beta Y + Z$, $z = \beta Z + bY$. This completes the proof, since $X \neq 0$. □

In 1950 the Canadian mathematician **Holzer** refined Legendre's theorem as follows (cf. [**C7**]). Holzer showed that if equation (1.8) has a nontrivial root, then it possesses a nontrivial solution $\{x, y, z\}$ such that

$$|x| \leq \sqrt{bc}, \quad |y| \leq \sqrt{ac}, \quad |z| \leq \sqrt{ab}.$$

This provides us with an effective algorithm for finding a nontrivial integer solution of equation (1.8) and, therefore, for finding a rational solution of equation (1.7).

5.1.3. THEOREM. *Let a, b and c be pairwise relatively prime square-free numbers such that $a, b > 0$ and $c < 0$. Then if $ax^2 + by^2 + cz^2 = 0$ has a nonzero integer solution, then it also has a nonzero integer solution for which*

$$x^2 \leq b|c|, \quad y^2 \leq a|c|, \quad z^2 \leq ab.$$

PROOF. (We follow Mordell [**C13**].) It suffices to prove that $z^2 \leq ab$. Indeed, then $ax^2 + by^2 \leq ab|c|$, hence, $x^2 \leq b|c|$ and $y^2 \leq a|c|$. Consider a solution $\{x_0, y_0, z_0\}$ for which $(x_0, y_0) = 1$ and $z_0^2 > ab$. It suffices to prove that with the help of this solution one can construct a new solution $\{x, y, z\}$ for which $|z| < |z_0|$.

The triple $x_0 + tx_1$, $y_0 + ty_1$, $z_0 + tz_1$ is a solution if and only if the numbers t, x_1, y_1, z_1 satisfy the relation

$$(ax_1^2 + by_1^2 + cz_1^2)t^2 + 2(ax_0x_1 + by_0y_1 + cz_0z_1)t = 0.$$

For $t \neq 0$ we can divide this relation by t and as a result we get $t = m/n$, where $m = -2(ax_0x_1 + by_0y_1 + cz_0z_1)$ and $n = ax_1^2 + by_1^2 + cz_1^2$. After multiplication by the denominator n we get a solution

$$x_0n + x_1m, \quad y_0n + y_1m, \quad z_0n + z_1m.$$

Let us show that these three numbers are divisible by $\Delta = (c, x_1y_0 - y_1x_0)$. Indeed, by the hypothesis $(c, ab) = 1$ and $(x_0, y_0) = 1$. Therefore, the identity $ax_0^2 + by_0^2 + cz_0^2 = 0$ implies that $(c, abx_0y_0) = 1$, hence, $(\Delta, abx_0y_0) = 1$. In particular, it is meaningful to speak about x_0^{-1} and y_0^{-1} modulo Δ, i.e., there exist integers x_0^{-1} and y_0^{-1} such that $x_0x_0^{-1} \equiv 1 \pmod{\Delta}$ and $y_0y_0^{-1} \equiv 1 \pmod{\Delta}$. Since $x_1y_0 - y_1x_0 \equiv 0 \pmod{\Delta}$, it follows that $x_1 \equiv y_1x_0y_0^{-1} \pmod{\Delta}$. Now it is easy to show that

$$P \equiv ax_0x_1 + by_0y_1 \equiv 0 \pmod{\Delta} \quad \text{and} \quad Q \equiv ax_1^2 + by_1^2 \equiv 0 \pmod{\Delta}.$$

Indeed,

$$P \equiv ax_0(y_1x_0y_1^{-1}) + by_0y_1 \equiv y_1(ax_0^2 + by_0^2)y_0^{-1} \equiv 0 \pmod{\Delta},$$
$$Q \equiv a(y_1x_0y_0^{-1})^2 + by_1^2 \equiv (ax_0^2 + by_0^2)y_1^2y_0^{-2} \equiv 0 \pmod{\Delta}.$$

Therefore,

$$\begin{aligned} x_0n + x_1m &= x_0(ax_1^2 + by_1^2 + cz_1^2) - 2x_1(ax_0x_1 + by_0y_1 + cz_0z_1) \\ &= x_0Q + cx_0z_1^2 - 2x_1P - 2cx_1z_0z_1 \equiv 0 \pmod{\Delta}. \end{aligned}$$

We similarly prove that the other two numbers are divisible by Δ.

Let δ be a divisor of Δ. Then the triple

$$x = (x_0 n + x_1 m)\delta^{-1}, \quad y = (y_0 n + y_1 m)\delta^{-1}, \quad z = (z_0 n + z_1 m)\delta^{-1}$$

is an integer solution. We have to select this solution so as to satisfy the inequality $|z| < |z_0|$. Clearly,

$$\frac{-\delta z}{c z_0} = \left(z_1 + \frac{a x_0 x_1 + b y_0 y_1}{c z_0}\right)^2 - \left(\frac{a x_0 x_1 + b y_0 y_1}{c z_0}\right)^2 - \frac{a x_1^2 + b y_1^2}{c}.$$

Taking into account that $c z_0^2 = -(a x_0^2 + b y_0^2)$ we get

$$\text{(1.11)} \qquad \frac{-\delta z}{c z_0} = \left(z_1 + \frac{a x_0 x_1 + b y_0 y_1}{c z_0}\right)^2 + \frac{ab}{c^2 z_0^2}(y_0 x_1 - x_0 y_1)^2.$$

Now we are ready to construct the solution required. For x_1 and y_1 take an arbitrary solution of the equation $y_0 x_1 - x_0 y_1 = \delta$. Then only one condition is imposed on δ, namely, δ divides c.

Case 1: c is even. Set $\delta = \frac{1}{2}c$ and select z_1 so that

$$\left| z_1 + \frac{a x_0 x_1 + b y_0 y_1}{c z_0} \right| \le \frac{1}{2}.$$

Then (1.11) yields

$$\frac{1}{2}\left|\frac{z}{z_0}\right| \le \frac{1}{4} + \frac{ab}{4 z_0^2}.$$

By the assumption, $z_0^2 > ab$, hence, $|z| < |z_0|$.

Case 2: c is odd. In this case we require the condition

$$\text{(1.12)} \qquad a x_1 + b y_1 + c z_1 \equiv 0 \pmod 2.$$

This condition corresponds to the choice of a definite parity of z_1. Relation (1.12) is equivalent to

$$n = a x_1^2 + b y_1^2 + c z_1^2 \equiv 0 \pmod 2.$$

Therefore, the numbers $x_0 n + x_1 m$, $y_0 n + y_1 m$, $z_0 n + z_1 m$ are divisible by 2δ. Hence, if instead of δ we take $\delta' = 2\delta$ it will also satisfy (1.11), which in this case takes the form

$$\frac{-2\delta z}{c z_0} = \left(z_1 + \frac{a x_0 x_1 + b y_0 y_1}{c z_0}\right)^2 + \frac{ab}{c^2 z_0^2}(y_0 x_1 - x_0 y_1)^2.$$

Set $\delta = c$ and select z_1 of the parity required (i.e., such that relation (1.12) holds) and such that

$$\left| z_1 + \frac{a x_0 x_1 + b y_0 y_1}{c z_0} \right| \le 1.$$

Then

$$2\left|\frac{z}{z_0}\right| \le 1 + \frac{ab}{z_0^2} < 2.$$

i.e., $|z| < |z_0|$. □

Problems

5.1.1. Which of the following equations have nontrivial integer solutions?
a) $3x^2 - 5y^2 + 7z^2 = 0$.
b) $7x^2 + 11y^2 - 19z^2 = 0$.
c) $8x^2 - 5y^2 - 3z^2 = 0$.
d) $11x^2 - 3y^2 - 41z^2 = 0$.

5.1.2. Find rational points of the following curves:
a) $x^2 - 3y^2 = 1$;
b) $x^2 + 2y^2 = 9$;
c) $x^2 - 6y^2 = 1$.

5.1.3. Prove that the equation $x^2 - 2y^2 = 3$ has no integer solutions.

5.1.4. Let d be a square-free natural number.

a) Prove that if (x_1, y_1) is an integer solution of equation $x^2 - dy^2 = 1$ and $(x_1 + y_1\sqrt{d})^n = x_n + y_n\sqrt{d}$, where x_n and y_n are integers, then (x_n, y_n) is also a solution of this equation.

b) Prove that if the equation $x^2 - dy^2 = -1$ has an integer solution, then the equation $x^2 - dy^2 = 1$ also has an integer solution.

c) Prove that equation $x^2 - dy^2 = 1$ has at least one integer solution.

d) Prove that if the equation $x^2 - dy^2 = n$, where $n \neq 0$, has at least one integer solution, then it has infinitely many of them.

§5.2. Addition of points on a cubic curve

In his *Arithmetics* Diophantus did not confine himself to the second degree equations. He succeeded in solving certain cubic equations as well and gave a general method for finding rational solutions of the equation

$$y(6 - y) = x^3 - x.$$

However, a steady interest in third degree diophantine equations first appeared, it seems, only in connection with a problem from antiquity on congruent numbers; the study of these numbers was initiated by Arab mathematicians in the tenth century.

A positive number $r \in \mathbb{Q}$ is called *congruent* if it is equal to the area of an acute triangle with sides of rational lengths. For example, 6 is the area of the triangle with sides 3, 4 and 5; hence, 6 is a congruent number. Let r be congruent and $a, b, c \in \mathbb{Q}$ be the lengths of the sides of an acute triangle with area r. For any $r \in \mathbb{Q}$ we can find $s \in \mathbb{Q}$ such that s^2r is a square-free integer. But the area of the triangle with sides sa, sb, and sc is equal to s^2r. Therefore, without loss of generality, we may assume that r is a square-free positive integer.

It is worth observing that in the definition of a congruent number the sides of the triangle can be rationals, not necessarily integers. Whereas 6 is the least possible area of the triangle with *integer* sides, it is possible to find an acute triangle of area 5 with *rational* sides. Indeed, the triangle with sides $\frac{3}{2}, \frac{20}{3}, \frac{41}{6}$ has such an area. One can prove that 5 is the smallest congruent integer.

It turns out that the problem of description of all congruent integers can be reduced to a third degree diophantine equation. Namely, the following statement holds.

5.2.1. Theorem. *Let n be a square-free natural number. Then the following three conditions are equivalent:*

(1) *n is a congruent number*;

(2) *there exists a rational number x such that the numbers x, $x+n$, and $x-n$ are squares of rational numbers*;

(3) *on the curve $y^2 = x^3 - n^2x$, there exists a rational point (x, y) whose coordinate x is the square of a rational number such that the denominator of x is an even number and the numerator of x has no common divisors with n.*

PROOF. Let $a < b < c$ be a triple of positive rational numbers such that $a^2+b^2 = c^2$ and $n = \frac{1}{2}ab$. Set $x = \frac{1}{4}c^2$. Then $x+n = \frac{1}{4}(a+b)^2$ and $x-n = \frac{1}{4}(a-b)^2$. Hence, x, $x+n$ and $x-n$ are the squares of rational numbers. On the other hand, if x satisfies condition (2), then set $c = 2\sqrt{x}$ and find a and b from the system of equations

$$\begin{cases} (a+b)^2 = 4(x+n), \\ (a-b)^2 = 4(x-n). \end{cases}$$

In other words, if x satisfies condition (2), then the desired triangle has sides of lengths

$$a = \sqrt{x+n} - \sqrt{x-n}, \quad b = \sqrt{x+n} + \sqrt{x-n}, \quad c = 2\sqrt{x}.$$

In order to prove the equivalence of (2) and (3), let us consider a rational number x such that $x, x+n$ and $x-n$ are squares of rational numbers. Then $x = u^2$ and $(x+n)(x-n) = u^4 - n^2 = v^2$. Set $y = uv$. Then

$$y^2 = x^3 - n^2x, \tag{2.1}$$

i.e., the point (x, y) belongs to the cubic (2.1). Since $x = \frac{1}{4}c^2$, the denominator of x is divisible by 2. Moreover, it is clear that the numerator of x has no common divisors with n.

Conversely, if $x = u^2 = (c/2)^2$ and $x^3 - n^2x = y^2$, then

$$v^2 = y^2/x = x^2 - n^2 = (x+n)(x-n)$$

and we have a Pythagorean triple $v^2 + n^2 = x^2$. The numbers x^2 and $v^2 = x^2 - n^2$ have the same denominator q^4 and the number q is even by the assumption. Hence, the numbers q^2v and q^2x are integers and q^2n is an even number such that q^2x and q^2n have no common divisors and

$$(q^2v)^2 + (q^2n)^2 = (q^2x)^2.$$

Hence, $q^2v = s^2 - t^2$, $q^2n = 2st$ and $q^2x = s^2 + t^2$, where s and t are integers. Since

$$(2s/q)^2 + (2t/q)^2 = 4x = (2u)^2,$$

the triangle with sides $2s/q$, $2t/q$ and $2u$ is a right one and its area is equal to $2st/q^2 = q^2n/q^2 = n$.

Theorem 5.2.1 is completely proved. □

The problem on congruent numbers and also certain other classical problems, for example, the problem of finding rational solutions of the equation $x^3 + y^3 = 1$, are particular cases of the problem of finding rational solutions of the general cubic equation $f(x, y) = 0$ for two unknowns, i.e., of finding rational points on the curve C defined by the equation $f(x, y) = 0$.

First, suppose that the cubic $f(x, y) = 0$ has a singular point O and that this point is rational. Any line passing through a singular point intersects the curve with multiplicity at least two. This implies, in particular, that a cubic curve cannot have two singular points, O and O_1, because otherwise the straight line OO_1 would intersect the cubic with multiplicity at least 4.

Let us draw a rational straight line $x = x_0 + at, y = y_0 + bt$, $a, b \in \mathbb{Q}$, through the point $O = (x_0, y_0)$. The intersection points of this line with the cubic correspond to the roots of the polynomial $F(t) = f(x_0 + at, y_0 + bt)$. The coefficients of F are rational and for almost all lines the degree of F is equal to 3 (the degree is smaller than 3 only for the lines passing through infinite points of the curve). The polynomial F has the root $t = 0$ of multiplicity 2 corresponding to point O. Hence, the third root of F is rational, i.e., it corresponds to a rational point on the curve. It is also clear that the straight line that connects O with a rational point on the curve is rational. This gives a complete description of the set of rational points of a singular cubic curve.

In what follows we will only consider nonsingular cubic curves. Recall that in Chapter 1 we defined addition of points on a nonsingular cubic curves as follows. Let E be a fixed point on the given curve, A and B points on the curve and X the intersection point of the straight line AB with the curve. The sum of points A and B is the intersection point of the straight line EX with the given curve. It is easy to verify that if the cubic curve is given by a polynomial with integer coefficients and the point E is rational, then the sum of two rational points is a rational point.

The curve $y^2 = x^3 + ax^2 + b$ is nonsingular if and only if its discriminant $\Delta = -(4a^3 + 27b^2)$ is nonzero. If we take the infinite point as the zero element, then for such a curve it is easy to get explicit formulas for addition of points. Let the straight line $y = px + q$ intersect the given curve at points (x_i, y_i), $i = 1, 2, 3$. Then

$$(px + q)^2 = x^3 + ax + b$$

for $x = x_1, x_2, x_3$. The sum of the roots of this equation is equal to p^2; hence,

$$x_3 = -x_1 - x_2 + p^2 = -x_1 - x_2 + \left(\frac{y_1 - y_2}{x_1 - x_2}\right)^2,$$

$$y_3 = px + q = \frac{y_1 - y_2}{x_1 - x_2}(x_3 - x_1) + y_1.$$

Clearly, the coordinates of the sum of the points (x_1, y_1) and (x_2, y_2) are $(x_3, -y_3)$. If $x_1 = x_2$ and $y_1 = y_2$, it suffices to notice that

$$\lim_{x_2 \to x_1} \frac{y_1 - y_2}{x_1 - x_2} = y'(x_1) = \frac{3x_1^2 + a}{2y_1}.$$

We should separately consider the case $x_1 = x_2$ and $y_1 \neq y_2$. In this case the sum of points is the infinite point of the curve.

The formulas obtained show that knowing one rational point P of the curve $y^2 = x^3 + ax + b$ we can find the rational points $2P$, $3P$, etc. Consider, for example, the curve $y^2 = x^3 - 2$ and point $P = (3, 5)$. Then

$$2P = \left(\frac{129}{100}, -\frac{383}{1000}\right)$$

is a new rational point. We can now compute $3P$, $4P$, etc. With every step the volume of calculations steeply increases. If we denote by x_n the first coordinate of the point nP, then

$$x_1 = 3, \quad x_2 = \frac{129}{100}, \quad x_3 = \frac{164323}{29241},$$
$$x_4 = \frac{2340922881}{58675600}, \quad x_5 = \frac{307326105747363}{160280942564521}.$$

The numerator of x_{11} has 71 digits.

It is worth observing that the points P, $2P$, $3P$, etc. are not necessarily distinct. If some of them coincide, then the least number $m \in \mathbb{Q}$ for which mP is the zero element of the group, i.e., the infinite point, is called *the order* of the point P.

Problems

5.2.1. Prove that the point $P = (0, 2)$ on the curve $y^2 = x^3 + 4$ is of order 3.
5.2.2. Prove that the point $P = (2, 4)$ on the curve $y^2 = x^3 + 4x$ is of order 4.

Each of the points in Problems 5.2.3 – 5.2.9 is of finite order on the corresponding curve. Find this order.
5.2.3. $P = (0, 4)$ on $y^2 = 4x^3 + 16$.
5.2.4. $P = (2, 8)$ on $y^2 = 4x^3 + 16x$.
5.2.5. $P = (2, 3)$ on $y^2 = x^3 + 1$.
5.2.6. $P = (3, 8)$ on $y^2 = x^3 - 43x + 166$.
5.2.7. $P = (3, 12)$ on $y^2 = x^3 - 14x^2 + 81x$.
5.2.8. $P = (0, 0)$ on $y^2 + y = x^3 - x^2$.
5.2.9. $P = (1, 0)$ on $y^2 + xy + y = x^3 - x^2 - 3x + 3$.
5.2.10. Let f_n be the functions on the curve

$$y^2 = x^3 + ax + b, \quad \text{where} \quad 4a^3 + 27b^2 \neq 0,$$

determined by the relations

$$\begin{aligned}
f_1 &= 1,\\
f_2 &= 2y,\\
f_3 &= 3x^4 + 6ax^2 + 12bx - a^2,\\
f_4 &= 4y(x^6 + 5ax^4 + 20bx^3 - 5a^2x^2 - 4abx - 8b^2 - a^3),\\
f_{2m} &= 2f_m(f_{m+2}f_{m-1} - f_{m+2}^2 f_{m+1}^2) \text{ for } m \geq 3,\\
f_{2m+1} &= f_{m+2}f_m^3 - f_{m-1}f_{m+1}^3 \text{ for } m \geq 2.
\end{aligned}$$

Further, let

$$g_n = xf_n^2 - f_{n-1}f_{n+1} \quad \text{and} \quad 4yh_n = f_{n+2}f_{n-1}^2 - f_{n-1}f_{n+1}^2.$$

Prove that

$$n(x, y) = \left(\frac{g_n}{f_n^2}, \frac{h_n}{f_n^3}\right).$$

§5.3. Several examples

In this section we consider several examples of nonsingular cubic curves determined by the equation in the normal form

$$y^2 = x^3 + ax + b$$

or by the equation

$$y^2 + 2cy = x^3 + ax + b$$

which can be reduced to the preceding one by the change of variables $y \mapsto y + c$.

We will be mainly interested in the set of rational points on the curve E; it will be denoted by $E(\mathbb{Q})$. This set, as we have already said, is an abelian group whose zero element is the infinite point on the curve. This is why it is also convenient to consider the infinite point of the curve as a rational point. The first three examples are taken from [**B10**].

EXAMPLE 1. Consider the following problem: *Represent the product of two consecutive integers* $y(y+1)$ *in the form of the product of three consecutive integers* $(x-1)x(x+1) = x^3 - x$. This problem leads to the curve E determined by the equation

$$y^2 + y = x^3 - x. \tag{3.1}$$

On this curve, there are six obvious points with integer coordinates (cf. Figure 39):

$$(0,0),\ (1,0),\ (-1,0),\ (0,-1),\ (1,-1),\ \text{and } (-1,-1).$$

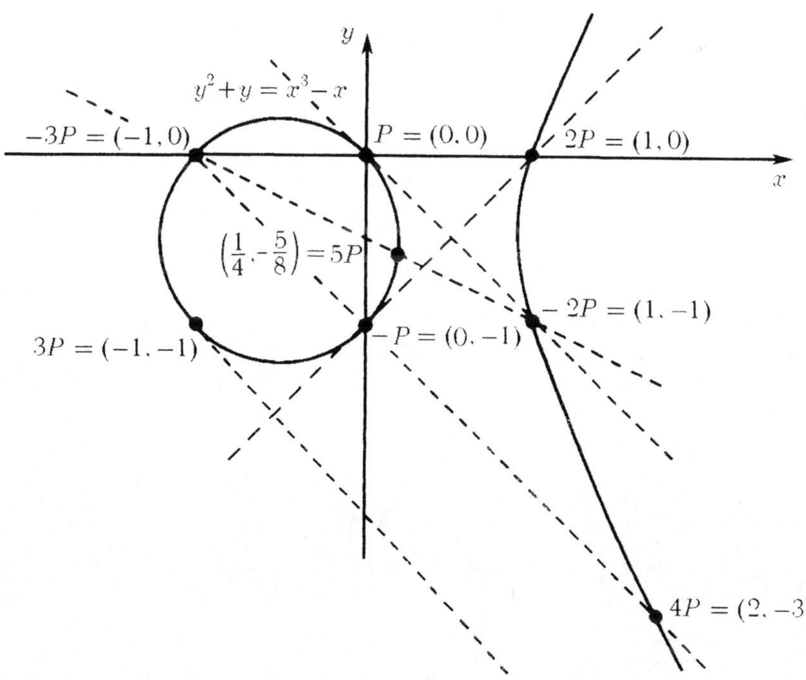

FIGURE 39

Set $P = (0,0)$. Then all the points indicated are generated by the point P:

$$(1,0) = 2P,\ (-1,0) = -3P,\ (0,-1) = -P,\ (1,-1) = -2P,\ (-1,-1) = 3P.$$

The point P actually generates an infinite cyclic group.

All the points of the form $(2n+1)P$ belong to the closed component of the curve containing P; the points of the form $2nP$ lie on the noncompact component and tend to infinity as n grows.

EXAMPLE 2. The curve E given by equation

$$y^2 + y = x^3 - x^2 \tag{3.2}$$

has four obvious points with integer coordinates (cf. Figure 40):

$$(1,0), (0,0), (0,-1) = -(0,0), \text{ and } (1,-1) = -(1,0).$$

The tangent to the curve E at $(1,0)$ intersects E at the point $(0,-1)$; this means that $2(1,0) = (0,0)$; hence, $2(1,-1) = (0,-1)$.

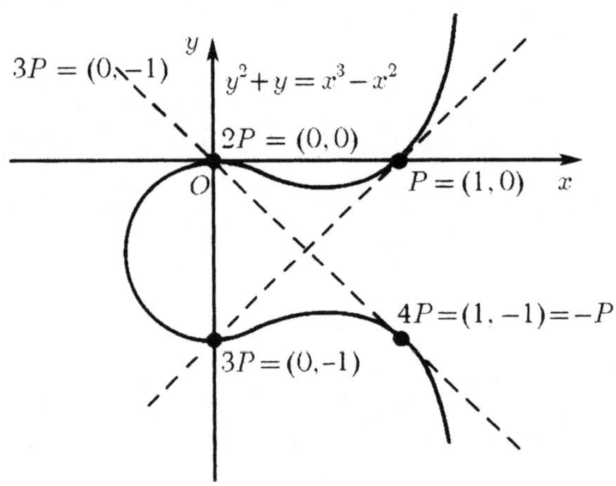

FIGURES 40

The tangent to E at the point $(0,0)$ intersects E at the point $(1,0)$; this means that $2(0,0) = (1,-1)$. The equations $2(1,0) = (0,0)$, $2(0,0) = (1,-1) = -(1,0)$ imply that $4(1,0) = (1,-1) = -(1,0)$, i.e., $5(1,0) = 0$. Hence, the subset

$$\{0,\ (1,0),\ (0,0),\ (0,-1),\ (1,-1)\}$$

is a cyclic subgroup of order 5 in $E(\mathbb{Q})$. Using a more advanced technique one can show that there are no other rational points on this curve.

EXAMPLE 3. Consider the curve E given by equation

$$y^2 + y = x^3 + x^2. \tag{3.3}$$

This curve has four obvious points with integer coordinates (cf. Figure 41):

$$(0,0),\ (-1,0),\ (0,-1) = -(0,0) \text{ and } (-1,-1) = -(-1,0).$$

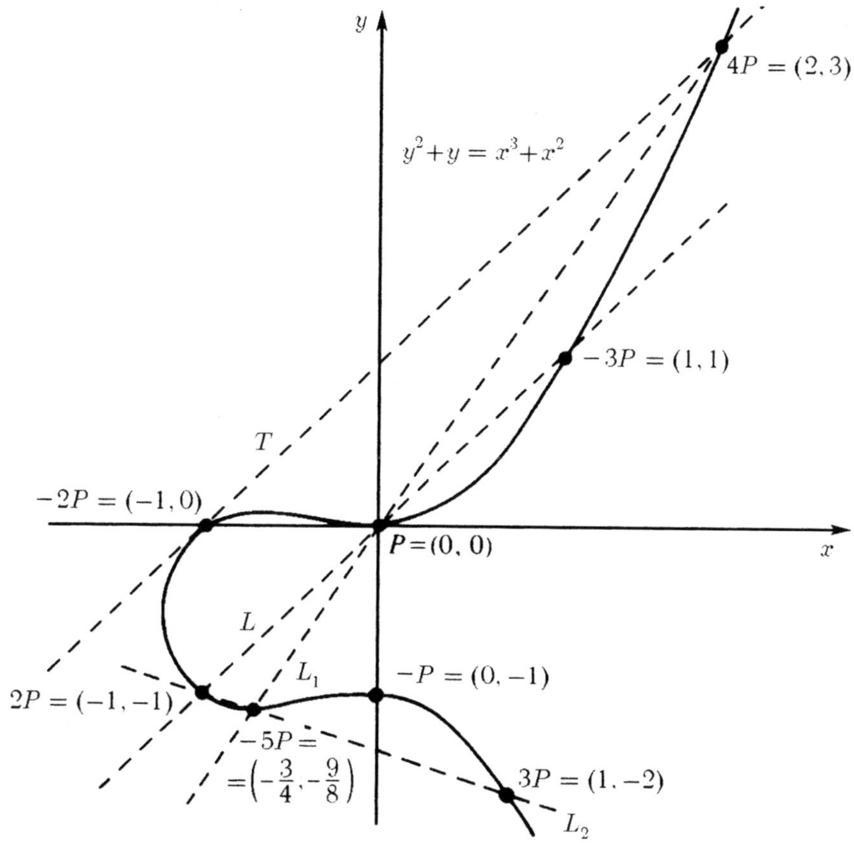

FIGURE 41

The point $P = (0,0)$ generates an infinite cyclic subgroup in $E(\mathbb{Q})$.

For instance, it is not difficult to calculate that

$$2P = (-1,-1), \quad -3P = (1,1), \quad 3P = (1,-2), \quad 4P = (2,3), \quad 5P = \left(-\frac{3}{4}, -\frac{9}{8}\right).$$

The tangent T to the curve at the point $-2P$ intersects the curve at the point $4P$ and the line L passing through $2P$ and P intersects E at the point $-3P$; the point $-5P$ is constructed either using the straight line L_1 through P and $4P$ or using the straight line L_2 through $2P$ and $3P$.

EXAMPLE 4. The curve E of this example is given by Fermat's equation

$$u^3 + v^3 = w^3. \tag{3.4}$$

The triples

$$(u, v, w) = (1, -1, 0), \quad (1, 0, 1) \text{ and } (0, 1, 1)$$

are nonzero integer solutions of this equation. All the other integer solutions are proportional to these solutions. The cubic equation (3.4) is not given in its normal form; however, the transformation

$$x = \frac{3w}{u+v}, \quad y = \frac{9}{2}\left(\frac{u-v}{u+v}\right) + \frac{1}{2} \tag{3.5}$$

reduces E, or rather, its equation, to the normal form:

$$y^2 - y = x^3 - 7. \tag{3.6}$$

The transformation (3.5) sends the point $(1, -1, 0)$ into the infinite point; the point $(1, 0, 1)$ into $(3, 5)$ and the point $(0, 1, 1)$ into $(3, -4)$. Thus, $E(\mathbb{Q})$ is a cyclic group consisting of three elements $\{0, (3,5), (3,-4)\}$ because if $E(\mathbb{Q})$ had contained other rational points (x, y), then there would have existed nontrivial solutions (u, v, w) of Fermat's equation $u^3 + v^3 = w^3$ corresponding to them.

Let us establish that $3P = 0$ for $P = (3,5)$ by computing $-2P$ using the tangent to E at point P. From the equation for the tangent

$$(2y - 1)y' = 3x^2$$

we find the coefficient of the slope of the tangent to E at the point $(3, 5)$:

$$\lambda = \frac{3x^2}{2y-1} = \frac{27}{9} = 3.$$

Hence, the tangent is given by the equation

$$y = 5 + 3(x - 3) = 3x - 4.$$

The tangent intersects the curve (3.6) at the points whose abscissas satisfy the equation

$$(3x - 4)^2 - (3x - 4) = x^3 - 7,$$

i.e.,

$$x^3 - 9x^2 + 27x - 27 = (x - 3)^3 = 0.$$

Thus, the abscissa of the point $-2P$ is $x = 3$. Hence, $-2P = P$, and so we have $3P = 0$.

EXAMPLE 5. Consider the curve

$$y^2 = x^3 + k, \tag{3.7}$$

where k is an integer.

The corresponding diophantine equation was first considered in the seventeenth century by **Fermat** and **Bashe de Mesiriac** in the particular case $k = -2$ and was later the subject of intensive study. It is as yet unknown for which integers k equation (3.7) has at least one rational solution. Bashe claimed (without proof) that if a rational solution (x, y) with $xy \neq 0$ exists, then the method of tangents leads to an infinite number of rational solutions.

In modern terms this means that if the group $E(\mathbb{Q})$ of rational points of the curve (3.7) is nonzero, then it is infinite. With certain restrictions this statement was proved in 1930 by the German mathematician **Fueter**. In 1966 a remarkably short proof of Fueter's result was discovered by the English mathematician **Mordell** (see Problem 5.3.2 below).

PROBLEMS

5.3.1. Let $P_0 = (x_0, y_0)$, where $y_0 \neq 0$, be a rational point of the curve

$$y^2 = x^3 + k.$$

Show that the tangent at P_0 intersects this curve at the point $P_1 = (x_1, y_1)$, where

$$x_1 = \frac{9x_0^4 - 8x_0y^2}{4y_0^2}, \qquad y_1 = \frac{27x_0^6 - 36x_0^3y_0^2 + 8y_0^4}{8y_0^3}.$$

In particular, show that the curve $y^2 = x^3 - 2$ has rational points $P_0 = (3, 5)$ and $P_1 = \left(\frac{129}{100}, \frac{383}{1000}\right)$.

5.3.2. (Mordell, [**C12**]) Let k be an integer free of 6th powers and distinct from 1 and -432. Further, let the curve

$$y^2 = x^3 + k$$

have a rational point $P_0 = (x_0, y_0)$, where $x_0y_0 \neq 0$.

a) Set $x_0 = \frac{p}{q^2}$ and $y_0 = \frac{r}{q^3}$, where $(p, q, r) = 1$, $(p, q) = 1$ and $(q, r) = 1$. Show that the coordinate x_1 of the point $P_1 = (x_1, y_1)$ from Problem 5.3.1 is given by the relation

$$q^2x_1 = \frac{9p^4}{4r^2} - 2p.$$

b) Prove that if $\frac{3p^2}{2r}$ is not an integer, then $P_1 \neq P_0$.

c) Prove that if $\frac{3p^2}{2r}$ is an integer, then, applying the process indicated in the preceding problem to the point P_1, we get a rational point $P_1' \neq P_0$.

d) Using the results of a)–c) prove that the curve $y^2 = x^3 + k$ has infinitely many rational points.

5.3.3. a) Prove that the equation $y^2 = x^3 + 1$ has no rational solutions distinct from $(-1, 0)$, $(0, \pm 1)$, $(2, \pm 3)$.

b) Prove that equation $y^2 = x^3 - 432$ has no rational solutions distinct from the solution $(12, \pm 36)$.

§5.4. Mordell's theorem

In 1901 the great French mathematician **A. Poincaré** conjectured [**A13**] that all rational points $E(\mathbb{Q})$ of an elliptic curve can be obtained as sums of a finite number of points. In algebraic terms, this statement can be formulated as follows.

5.4.1. THEOREM. *The group $E(\mathbb{Q})$ of rational points of any rational elliptic curve E is a finitely generated abelian group.*

Poincaré himself considered this statement obvious. In 1922 the English mathematician **Mordell** [**C11**] obtained the first rigorous proof of Poincaré's conjecture. During the following seven decades various generalizations and new variants of the proof of this theorem appeared. We will give one of them here. Our proof has a small gap; in order to fill in the gap we need the machinery of the theory of divisibility in rings of integer algebraic numbers; this goes beyond the scope of our book (see [**A13**]).

Consider the curve

$$y^2 = x^3 + ax + b, \tag{4.0}$$

where $a, b \in \mathbb{Q}$ and $\Delta = -(4a^3 + 27b^2) \neq 0$. We may assume that a and b are not just rational but integer numbers. Indeed, let

$$a = \frac{p}{q}, \quad b = \frac{r}{q}$$

(we do not suppose that the fractions are irreducible). Set

$$y = \frac{y_1}{q^3}, \quad x = \frac{x_1}{q^2}.$$

Then the initial equation takes the form (with integer coefficients in the right-hand side):

$$y_1^2 = x_1^3 + pq^3 x_1 + rq^5.$$

Let α_1, α_2, α_3 be the roots of equation (4.0). Since by assumption $a, b \in \mathbb{Z}$, it follows that α_1, α_2, α_3 are integer algebraic numbers. Thus,

$$y^2 = (x - \alpha_1)(x - \alpha_2)(x - \alpha_3). \tag{4.1}$$

It suffices to keep track of the coordinate x only since two values $\pm y_0$ correspond to every value x_0. Let us consider, for example, the elliptic curve $y^2 = x^3 - 2$. The point $P_0 = (3, 5)$ lies on this curve. In the preceding section we have shown that

$$2P_0 = \left(\frac{129}{100}, -\frac{383}{1000}\right).$$

Moreover,

$$x_1 - \alpha_1 = \frac{129}{100} - \sqrt[3]{2} = \left(\frac{9 - 6\sqrt[3]{2} - 2\sqrt[3]{4}}{10}\right)^2,$$

i.e., the point $2P_0 = (x_1, y_1)$ has the following property: the number $x_1 - \alpha_1$ is a perfect square in the field $\mathbb{Q}(\sqrt[3]{2})$. This property plays a decisive role in the proof of Mordell's theorem. Let us show that if E is a rational elliptic curve (4.1) and α is one of the integer algebraic numbers α_1, α_2, α_3, then the map $E(\mathbb{Q}) \longrightarrow \mathbb{Q}(\alpha)$ that sends the point $(x_0, y_0) \in E(\mathbb{Q})$ into $x_0 - \alpha \in \mathbb{Q}(\alpha)$, can be included in the commutative diagram (the proof of commutativity will be given shortly)

$$\begin{array}{ccc} E(\mathbb{Q}) & \longrightarrow & \mathbb{Q}(\alpha) \\ \downarrow & & \downarrow \\ E(\mathbb{Q})/2E(\mathbb{Q}) & \longrightarrow & \mathbb{Q}(\alpha)/\mathbb{Q}(\alpha)^2 \end{array}$$

in which the addition of points in $E(\mathbb{Q})$ corresponds to the multiplication of cosets in $\mathbb{Q}(\alpha)/\mathbb{Q}(\alpha)^2$.

The fact that the image of the set $E(\mathbb{Q})$ in $\mathbb{Q}(\alpha)/\mathbb{Q}(\alpha)^2$ is finite (we will not prove this fact) is of extreme importance.

Let us first obtain explicit formulas for addition of points on the curve $y^2 = x^3 + ax + b$ in the form most convenient for our goals. Let the straight line $y = px + q$ intersect the curve $y^2 = x^3 + ax + b$ at three points (x_i, y_i), $i = 1, 2, 3$. Then

$$x^3 + ax + b - (px + q) = (x - x_1)(x - x_2)(x - x_3).$$

In particular, if $x = \alpha$ is one of the roots of the polynomial $x^3 + ax + b$, then taking into account that $\alpha^3 + a\alpha + b = 0$, we get

$$(x_1 - \alpha)(x_2 - \alpha)(x_3 - \alpha) = (p\alpha + q)^2.$$

It is also clear that

$$p\alpha + q = \frac{y_1(x_2 - \alpha) - y_2(x_1 - \alpha)}{x_2 - x_1}.$$

Therefore,

$$x_3 - \alpha = \frac{1}{(x_1 - \alpha)(x_2 - \alpha)} \left(\frac{y_1(x_2 - \alpha) - y_2(x_1 - \alpha)}{x_2 - x_1} \right)^2. \tag{4.2}$$

If $x_1 = x_2$ (i.e., for the tangent) we get

$$p = \frac{3x^2 + a}{2y_1}.$$

Hence,

$$x_3 - \alpha = \frac{1}{(x_1 - \alpha)^2} \left(\frac{3x^2 + a}{2y_1}(\alpha - x_1) + y_1 \right)^2 = \left(\frac{1}{2y_1} \left(3x^2 + a + \frac{2y_1^2}{\alpha - x_1} \right) \right)^2. \tag{4.3}$$

If we make use of the relations

$$y_1^2 = (x_1 - \alpha)(x_1 - \alpha_2)(x_1 - \alpha_3),$$
$$\alpha + \alpha_2 + \alpha_3 = 0, \quad \alpha\alpha_2\alpha_3 = -b, \quad \alpha^3 + a\alpha + b = 0,$$

then (4.3) can be reduced to the form

$$x_3 - \alpha = \left(\frac{3x_1^2 - a - 2\alpha x_1 - 2\alpha^2}{2y_1} \right)^2. \tag{4.4}$$

Now we are ready to prove the commutativity of the above diagram. In the group $\mathbb{Q}(\alpha)/\mathbb{Q}(\alpha)^2$ the order of any nonunit elements is equal to 2. Therefore, formula (4.2) shows that *if the classes $A, B \in \mathbb{Q}(\alpha)/\mathbb{Q}(\alpha)^2$ correspond to the points P and Q, then the class AB corresponds to the point $P + Q$*. Formula (4.4) implies that *the unit class E corresponds to the point $2P$*.

This statement can be inverted: if the point Q corresponds to the unit class, then $Q = 2P$. In proving this fact we will confine ourselves to the case when the elements 1, α and α^2 are linearly independent over $\mathbb{Q}$. Let

$$\begin{aligned} x - \alpha &= (u_0 + u_1\alpha + u_2\alpha^2)^2 \\ &= u_0^2 + 2u_0u_1\alpha + (u_1^2 + 2u_0u_1)\alpha^2 + 2u_1u_2\alpha^3 + u_2^2\alpha^4 \\ &= (u_0^2 - 2bu_1u_2) + (2u_0u_1 - 2au_1u_2 - bu_2^2) + (u_1 + 2u_0u_2 - au_2^2)\alpha^2. \end{aligned}$$

Then

(i) $u_0^2 - 2bu_1u_2 = x$,
(ii) $2u_0u_1 - 2au_1u_2 - bu_2^2 = -1$,
(iii) $u_1 + 2u_0u_2 - au_2^2 = 0$.

Let us multiply (ii) and (iii) by $-u_2$ and u_1, respectively, and add the equations obtained. We get

$$u_1^3 + au_1u_2^2 + bu_2^2 = u_2.$$

Since $u_2 \neq 0$, it follows that

$$\frac{1}{u_2^2} = \left(\frac{u_1}{u_2}\right)^3 + a\frac{u_1}{u_2} + b,$$

i.e., the point $P = (x', y')$, where $x' = \dfrac{u_1}{u_2}$ and $y' = \dfrac{1}{u_2}$, lies on the curve considered.

Let us prove that $Q = 2P$. Relation (iii) implies that

$$x'^2 + 2u_0y' - a = 0,$$

hence, $u_0 = \dfrac{a - x'^2}{2y'}$. Moreover, $u_2 = \dfrac{1}{2y'}$, $u_1 = x'u_2 = \dfrac{x'}{y'}$. Therefore,

$$u_0 + u_1\alpha + u_2\alpha^2 = \frac{a - x'^2}{2y'} + \frac{x'}{y'}\alpha + \frac{1}{y'}\alpha^2 = -\frac{x'^2 - \alpha - 2\alpha x' - 2\alpha^2}{2y'}.$$

By (4.4), this implies that $Q = 2P$. Therefore, if the points P and Q correspond to one class, then $P + Q$ corresponds to the class E and, therefore, there exists a point R such that $P + Q = 2R$. This completes the proof of commutativity of our diagram.

As we have noted above, one can prove that the image of the group $E(\mathbb{Q})$ in $\mathbb{Q}(\alpha)/\mathbb{Q}(\alpha)^2$ is finite. (We skip the proof in order to avoid a rather long excursion into the theory of divisibility in rings of integer algebraic numbers.) Therefore, the rational points of our curve belong to a finite number of classes from $\mathbb{Q}(\alpha)/\mathbb{Q}(\alpha)^2$; let the points $Q_1, \dots, Q_m$ be representatives of these classes. Let us take an arbitrary rational point P_0 on the curve. This point belongs to the same class as one of the points Q_{i_1}; hence, $P_0 + Q_{i_1} = 2P_1$. Similarly, $P_1 + Q_{i_2} = 2P_2$. Therefore,

$$P_0 + Q_{i_1} + 2Q_{i_2} = 2(P_1 + Q_{i_2}) = 4P_2.$$

By continuing similar arguments we find that

$$P_0 + Q_{i_1} + 2Q_{i_2} + \dots + 2^kQ_{i_{k+1}} = 2^{k+1}P_{k+1}. \tag{4.5}$$

In what follows we will show that for any point P_0 after finitely many steps we get a point P_{k+1}, for which the numerators and denominators of its coordinates are bounded by a constant C independent of P_0. There are finitely many such points and, therefore, any point P_0 belongs to the group generated by the union of the points $Q_1, \dots, Q_m$ with a finite set of points. This is precisely the statement of Mordell's theorem.

Thus, it remains to prove that the procedure (4.5) leads to a point P_{k+1} of the type desired.

We can represent any point (x_0, y_0) on the curve $y^2 = x^3 + ax + b$ in the form $(\frac{p}{s^2}, \frac{t}{s^3})$, where $\frac{p}{s^2}$ and $\frac{t}{s^3}$ are irreducible fractions. Indeed, let $x_0 = \frac{p}{q}$ be an irreducible fraction. Then

$$(y_0q^2)^2 = q(p^3 + apq^2 + bq^3) = qr,$$

where $(q, r) = (q, p^3) = 1$. Therefore, $q = s^2$, $r = t^2$ and $y_0^2s^8 = s^2t^2$, i.e., $y = \frac{t}{s^3}$. Since $(p, q) = 1$ and $(q, r) = 1$, it follows that $(p, s) = 1$ and $(t, s) = 1$.

Let $\lambda_0 = \max(|p|, s^2)$. Then $t^2 \leq \lambda_0^3(1 + |a| + |b|)$, i.e., $|t| \leq c_1\lambda_0^{3/2}$. Thus, for the point $P_0 = (x_0, y_0)$ we have determined the number λ_0.

Similarly, let us determine the numbers ρ, λ_1 and ω for the points Q_{i_1}, P_1 and $2P_1$, respectively. We would like to estimate λ_1 in terms of λ_0. First, let us estimate ω. For convenience, designate the coordinates of the points as follows:

$$P_0 = \left(\frac{x}{z^2}, \frac{y}{z^3}\right), \quad Q_{i_1} = \left(\frac{p}{r^2}, \frac{q}{r^3}\right), \quad P_1 = \left(\frac{x_1}{z_1^2}, \frac{y_1}{z_1^3}\right), \quad 2P_1 = \left(\frac{s}{u^2}, \frac{t}{u^3}\right).$$

Here the numbers p, q, r can be assumed to be bounded by a constant (the numbers s, t, u are only used as an intermediary result of computations; the final result is the triple x_1, y_1, z_1).

The formula for addition of points on the cubic curve can be transformed to the following form:

$$x_3 = -x_1 - x_2 + \left(\frac{y_1 - y_2}{x_1 - x_2}\right)^2 = \frac{(x_1x_2 + a)(x_1 + x_2) + 2b - 2y_1y_2}{(x_1 - x_2)^2}.$$

Therefore, from the equation $P_0 + Q_{i_1} = 2P_1$ we derive

$$\frac{s}{u^2} = \frac{(px + ar^2z^2)(r^2x + pz^2) + 2br^4z^4 - 2qryz}{(r^2x - pz^2)^2}.$$

The orders of x and z^2 are both equal to λ_0, the order of z^4 is equal to λ_0^2, and the order of yz is equal to $\lambda_0^{3/2}\lambda_0^{1/2} = \lambda_0^2$; hence, $\omega \le c_2\lambda_0^2$.

Now we show that $\lambda_1 \le c_3\omega^{1/4} \le c_4\lambda_0^{1/2}$. Then $\lambda_2 \le c_4c_4^{1/2}\lambda_0^{1/4}$ and, therefore, $\lambda_n \le c\lambda_0^{1/2^n}$, where $c = \max(1, c_4^2)$. Hence, for any λ_0 and for sufficiently large n the number λ_n is bounded by a constant C. It is possible to express formula (4.4) in the form

$$\left(\frac{s}{u^2} - \alpha\right)^{1/2} = \left(\frac{x_1^2}{z_1^4} - a - 2\alpha\frac{x_1}{z_1^2} - 2\alpha^2\right)\frac{z_1^3}{2y_1},$$

i.e.,

$$\left(s - \alpha u^2\right)^{1/2} = \frac{u}{2y_1z_1}\left(x_1^2 - az_1^4 - 2\alpha x_1z_1^2 - 2\alpha^2z_1^4\right) = e_0 + e_1\alpha + e_2\alpha^2,$$

where

$$e_0 = \frac{u}{2y_1z_1}\left(x_1^2 - az_1^4\right), \quad e_1 = -\frac{ux_1z_1}{y_1}, \quad e_2 = \frac{uz_1^3}{y_1}.$$

The expression on the left-hand side is an algebraic integer; hence, $e_0 + e_1\alpha + e_2\alpha^2$ is also an algebraic integer.

5.4.2. LEMMA. *Let $e_0 + e_1\alpha + e_2\alpha^2$, where $e_i \in \mathbb{Q}$ and $\alpha^3 + a\alpha + b = 0$, be an integer algebraic number. Let Δ be the discriminant of the polynomial $x^3 + ax + b$. Then $\Delta e_i \in \mathbb{Z}$.*

PROOF. Let $\alpha_1, \alpha_2, \alpha_3$ be the roots of the polynomial $x^3 + ax + b$, and σ_1, σ_2, σ_3 be the automorphisms of $\mathbb{C}$ that send α into α_1, α_2, α_3, respectively. For $\gamma \in \mathbb{Q}(\alpha)$ define its *trace*, $\operatorname{tr}(\gamma)$, by setting

$$\operatorname{tr}(\gamma) = \sigma_1(\gamma) + \sigma_2(\gamma) + \sigma_3(\gamma).$$

Then $\operatorname{tr}(\gamma) \in \mathbb{Q}$ and if γ is an algebraic integer, then $\operatorname{tr}(\gamma) \in \mathbb{Z}$.

Let us multiply the equation $\gamma = e_0 + e_1\alpha + e_2\alpha^2$ by α and α^2 and take the traces. We get:

$$\begin{aligned} \mathrm{tr}(\gamma) &= 3e_0 + e_1 tr(\alpha) + e_2 tr(\alpha^2), \\ \mathrm{tr}(\alpha\gamma) &= e_0 tr(\alpha) + e_1 tr(\alpha^2) + e_2 tr(\alpha^3), \\ \mathrm{tr}(\alpha^2\gamma) &= e_0 tr(\alpha^2) + e_1 tr(\alpha^3) + e_2 tr(\alpha^4). \end{aligned}$$

These three equations can be considered as equations for e_0, e_1, e_2 (with integer coefficients). The determinant of this system is equal to

$$\begin{aligned} &\begin{vmatrix} 3 & \alpha_1+\alpha_2+\alpha_3 & \alpha_1^2+\alpha_2^2+\alpha_3^2 \\ \alpha_1+\alpha_2+\alpha_3 & \alpha_1^2+\alpha_2^2+\alpha_3^2 & \alpha_1^3+\alpha_2^3+\alpha_3^3 \\ \alpha_1^2+\alpha_2^2+\alpha_3^2 & \alpha_1^3+\alpha_2^3+\alpha_3^3 & \alpha_1^4+\alpha_2^4+\alpha_3^4 \end{vmatrix} \\ &\qquad = \begin{vmatrix} 1 & 1 & 1 \\ \alpha_1 & \alpha_2 & \alpha_3 \\ \alpha_1^2 & \alpha_2^2 & \alpha_3^2 \end{vmatrix} \cdot \begin{vmatrix} 1 & \alpha_1 & \alpha_1^2 \\ 1 & \alpha_2 & \alpha_2^2 \\ 1 & \alpha_3 & \alpha_3^2 \end{vmatrix} \\ &\qquad = (\alpha_1-\alpha_2)^2(\alpha_1-\alpha_3)^2(\alpha_2-\alpha_3)^2 = \Delta. \end{aligned}$$

Therefore, Lemma 5.4.2 is proved. □

Thus, the numbers $\Delta(2e_0 - ae_2) = \frac{\Delta u}{y_1 z_1}x_1^2$ and $\Delta e_2 = \frac{\Delta u}{y_1 z_1}z_1^4$ are integers. Since the fractions $\frac{x_1}{z_1^2}$ and $\frac{y_1}{z_1^3}$ are irreducible, it follows that $\frac{\Delta u}{y_1 z_1}$ is also an integer. Indeed, if the latter fraction were irreducible with the denominator divisible by y_1 (resp. z_1), then this factor could not cancel z_1^4 (resp. x_1^2). Therefore, both x_1^2 and z_1^4 are divisors of the numbers $\Delta(2e_0 - ae_2)$ and Δe_2.

Consider the equality (4.5) for the numbers α_1, α_2, and α_3. We see that $\Delta(2e_0 - ae_2)$ and Δe_2 can be linearly expressed in terms of $\left(s - \alpha_i u^2\right)^{1/2}$ for $i = 1$, 2, 3 and, therefore, they are of order $\omega^{1/2}$. Thus, we have shown that $\lambda_1 \le c_1\lambda_0^{1/2}$ and, in other words, we have established that the set of points P_{k+1} is finite, as required. □

§5.5. The rank and the torsion group of an elliptic curve

Consider an elliptic curve E determined over $\mathbb{Q}$. By the results of the preceding section the group $E(\mathbb{Q})$ of the rational points of E is a finitely generated abelian group. As any finitely generated abelian group, $E(\mathbb{Q})$ admits the following decomposition:

$$E(\mathbb{Q}) = \mathbb{Z}^{r_E} \times \mathrm{Tors}\, E(\mathbb{Q}),$$

where r_E is the rank of $E(\mathbb{Q})$ and $\mathrm{Tors}\, E(\mathbb{Q})$ is the subgroup of elements of finite order in $E(\mathbb{Q})$. The number r_E is called the *rank of the elliptic curve* E and the subgroup $\mathrm{Tors}\, E(\mathbb{Q})$ the *torsion group* of this elliptic curve. It is possible to prove that the rank of an elliptic curve is preserved under birational transformations.

The rank r_E is calculated for many elliptic curves over $\mathbb{Q}$. In the majority of cases it is rather small: most often it is equal to 0, 1, 2 or 3. As an illustration let us give the values of the ranks of the curves $y^2 = x^3 + ax$ and $y^2 = x^3 + a$ for several values of a (cf. [**B10**]).

Table 1. The ranks of elliptic curves E given by equation $y^2 = x^3 + ax$.

rank	the values of a
0	$1, 2, 4, 6, 7, 10, 11, 12, 22, -1, -3, -4, -8, -9, -11, -13, -18, -19$
1	$3, 5, 8, 9, 13, 15, 18, 19, 20, -2, -5, -6, -7, -10, -12, -14, -15, -20$
2	$14, 33, 34, 39, 46, -17, -56, -65, -77$
3	-82

Table 2. The ranks of elliptic curves E given by equation $y^2 = x^3 + a$.

rank	the values of a
0	$1, 4, 6, 7, 13, 14, 16, 20, 21, -1, -3, -5, -6, -8, -9, -10, -14, -432$
1	$2, 3, 5, 8, 9, 10, 11, 12, 18, -2, -4, -7, -13, -15, -18, -19, -20, -21$
2	$15, 17, 24, 37, 43, -11, -26, -39, -47$
3	$113, 141, 316, 346, 359, -174, -307, -362$

It is yet unknown if there exist elliptic curves of arbitrarily large rank. In 1986 the French mathematician **Mestre** constructed examples of elliptic curves of ranks 3 to 14. For example, the curve

$$y^2 + 9767y = x^3 + 3576x^2 + 425x - 2412$$

is of rank $r_E \geq 9$, and the curve

$$y^2 + 357573631y = x^3 + 2597055x^2 - 549082x - 19608054$$

is of rank $r_E \geq 14$.

One of the most famous conjectures in modern number theory relates the number r_E to the order of the zero at $s = 1$ of the analytic function corresponding to E. This conjecture was made in 1965 by the English mathematicians **Birch** and **Swinnerton–Dyer**. In order to formulate it, we need some prerequisites. Let

$$y^2 = x^3 + ax + b, \quad \Delta = -(4a^3 + 27b^2) \neq 0$$

be an elliptic curve. As we observed in the preceding section, we may assume without loss of generality that a and b are integers. Let p be a prime. Consider the congruence

$$y^2 \equiv x^3 + ax + b \pmod{p},$$

or, equivalently, the equation

$$y^2 = x^3 + \bar{a}x + \bar{b}, \quad \text{where} \quad \bar{a}, \bar{b} \in \mathbb{Z}/p\mathbb{Z} = \mathbb{F}_p. \tag{5.1}$$

If the prime p does not divide the discriminant Δ, then equation (5.1) determines an elliptic curve E_p over $\mathbb{F}_p$ called the *reduction of E modulo p*. Denote by N_p the number of points of E_p with coordinates in $\mathbb{F}_p$, the infinite point included.

For example, the solutions of the equation $y^2 = x^3 + 3x$ considered over $\mathbb{F}_5$ are

$$(0, 0),\ (1, \pm 2),\ (2, \pm 2),\ (3, \pm 1),\ (4, \pm 1),\ \infty;$$

hence, $N_5 = 10$.

What can one say about N_p for an arbitrary curve $y^2 = x^3 + \bar{a}x + \bar{b}$ over $\mathbb{F}_p$? Any such curve contains at least the infinite point. On the other hand, each element x from $\mathbb{F}_p$ produces not more than two values of y and, therefore, we see that N_p does not exceed $2p + 1$, counting the infinite point. Thus, we have the following obvious inequalities:

$$1 \leq N_p \leq 2p + 1.$$

These inequalities can be rewritten as follows:

$$|p + 1 - N_p| \leq p.$$

In 1934 the German mathematician **H. Hasse** obtained a finer estimate [**C6**]:

$$|p + 1 - N_p| \leq 2\sqrt{p}.$$

Starting from the numbers N_p, define the so-called *L-function of a rational elliptic curve E* by setting

$$L(E, s) = \prod_{p|\Delta} \left(\frac{1}{1 - a_p p^{-s}} \right) \cdot \prod_{p \nmid \Delta} \left(\frac{1}{1 - a_p p^{-s} + p^{1-2s}} \right),$$

where $a_p = p + 1 - N_p$ and Δ is the discriminant of the given curve.

From Hasse's estimate it is easy to derive that the above infinite product converges for $\operatorname{Re} s > \frac{3}{2}$. Conjecturally the function $L(E, s)$ can be analytically continued to the whole complex plane.

5.5.1. CONJECTURE 1 (Hasse–Weil). *For any rational elliptic curve E there exists a positive integer N and a sign $\varepsilon = \pm 1$ such that the modified L-function*

$$\Lambda(E, s) = N^{s/2}(2\pi)^{-s}\Gamma(s)L(E, s),$$

where $\Gamma(s)$ is the Euler Γ-function, satisfies the functional equation

$$\Lambda(E, s) = -\varepsilon\Lambda(E, 2 - s).$$

In spite of titanic efforts of many first-rate mathematicians, until the summer of 1993 this conjecture was only proved in certain particular cases. In June of 1993 a wonderful development occurred: Princeton University professor **A. Wiles** announced the proof of the Weil–Taniyama conjecture for semistable elliptic curves. The Weil–Taniyama conjecture implies the Hasse–Weil conjecture.

In this small book we cannot discuss Weil–Taniyama's conjecture; we will only observe that it states that every elliptic curve can be parameterized in a special way by the so-called modular functions. One can judge how deep the properties of the numbers mentioned in Weil–Taniyama's conjecture are from the fact that the conjecture implies, in particular, the proof of Fermat's Last Theorem.

Using vast empirical material on curves of the form $y^2 = x^3 + ax$ and $y^2 = x^3 + a$ and assuming that $L(E, s)$ can be continued to the whole complex plane, Birch and Swinnerton-Dyer came to the following conjecture.

5.5.2. CONJECTURE 2 (Birch–Swinnerton-Dyer). *Let E be a rational elliptic curve. Then its rank r_E is equal to the order of the zero of the function*

$$\tilde{L}(E, s) = \prod_{p \nmid \Delta} \frac{1}{1 - a_p p^{-s} + p^{1-2s}}$$

at point $s = 1$.

At the moment the Birch–Swinnerton-Dyer conjecture is only proved in certain particular cases.

Now we pass to the torsion group Tors $E(Q)$. To illustrate the possibilities that might occur, let us compute the torsion group for the family of curves $y^2 = x^3 + ax$. As before, by substituting $x \mapsto q^2x, y \mapsto q^3y$ we get for this family the equation

$$q^6y^2 = q^6\left(x^3 + \left(\frac{a}{q^4}\right)x\right).$$

This means that we may assume a to be a nonzero integer free of fourth powers.

5.5.3. THEOREM. *Let E be an elliptic curve determined by the equation $y^2 = x^3 + ax$, where a is an integer free of fourth powers. Then*

$$\operatorname{Tors} E(\mathbb{Q}) = \begin{cases} \mathbb{Z}/2 \oplus \mathbb{Z}/2 & \textit{if } a \textit{ is a perfect square distinct from } 4; \\ \mathbb{Z}/4 & \textit{if } a = 4; \\ \mathbb{Z}/2 & \textit{otherwise.} \end{cases}$$

PROOF. In all cases the order of the point $(0, 0)$ is equal to 2 since an arbitrary point of order 2 is of the form $(x, 0)$, where x is a root of the cubic equation $x^3 + ax = 0$. In particular, three points of order 2 exist if and only if $-a$ is a perfect square.

Now we consider the equation $2(x, y) = (0, 0)$. For such a point there exists a straight line $L : y = \lambda x$ passing through $(0, 0)$ and tangent to the curve E at the point (x, y). Therefore, $(\lambda x)^2 = x^3 + ax$ so that $x(x^2 - \lambda^2 x + a) = 0$. Since L is the tangent to E at the point (x, y), it follows that the quadratic equation $x^2 - \lambda^2 x + a = 0$ has a double root, i.e., its discriminant $\lambda^4 - 4a$ vanishes. Since a is free of fourth powers, the equation

$$\lambda^4 - 4a = 0$$

has a rational solution if and only if $a = 4$. In this case the points (x, y) satisfying the equation $2(x, y) = (0, 0)$ are of the form $(2, 4)$ and $(2, -4)$. Therefore, the points of order 2 and 4 in $E(\mathbb{Q})$ form, depending on a, subgroups of $\mathbb{Z}/2 \oplus \mathbb{Z}/2$, $\mathbb{Z}/4$ or $\mathbb{Z}/2$. So to prove the theorem, it suffices to show that there are no points of odd order in $E(\mathbb{Q})$.

We will carry out the proof of this fact by contradiction. Suppose that $E(\mathbb{Q})$ has a point $p \neq 0$ for which $3P = 0$, i.e., $2P = -P$. Then the tangent $y = \lambda x + \beta$ to E at P if substituted into the equation

$$y^2 = x^3 + ax$$

should give a perfect cube, i.e., the equation

$$0 = x^3 + ax - (\lambda x + \beta)^2 = (x - r)^3, \tag{5.2}$$

where r is the first coordinate of P, should be satisfied. By removing parentheses in relation (5.2) we get

$$\begin{aligned} 0 &= x^3 - \lambda^2 x^2 + (a - 2\beta\lambda)x - \beta^2 \\ &= x^3 - 3rx^2 + 3r^2x - r^3, \end{aligned}$$

from which $3r = \lambda^2$, $r^3 = \beta^2$ or, equivalently, $\beta^2 = \frac{\lambda^6}{27}$. Moreover, the third relation between the coefficients,

$$3r^2 = a - 2\beta\lambda,$$

leads to the equation

$$3\left(\frac{\lambda^4}{9}\right) = a - 2\left(\frac{\lambda^4}{3\sqrt{3}}\right),$$

which is impossible for rational a and λ. Therefore, Theorem 5.5.3 is proved. □

In a similar way we can prove that if E is the elliptic curve given by the equation $y^2 = x^3 + ax$, where a is a number free of sixth powers, then

$$\operatorname{Tors} E(\mathbb{Q}) = \begin{cases} \mathbb{Z}/6\mathbb{Z} & \text{if } a = 1; \\ \mathbb{Z}/3\mathbb{Z} & \text{if } a \text{ is a perfect square distinct from 1 or } a = -432; \\ \mathbb{Z}/2 & \text{if } a \text{ is a perfect cube distinct from 1;} \\ 0 & \text{otherwise.} \end{cases}$$

The structure of the group for the curves in the Weierstrass form $y^2 = f(x)$ with integer coefficients is considerably clarified by the following statement.

5.5.4. THEOREM. *Let E be the elliptic curve $y^2 = f(x)$, where*

$$f(x) = x^3 + ax^2 + bx + c$$

is a third degree polynomial with integer coefficients. If (x, y) is a point of finite order of E, then $x, y \in \mathbb{Z}$ and y divides the discriminant Δ of $f(x)$.

This theorem implies the existence of an effective computation algorithm for calculating the group $\operatorname{Tors} E(\mathbb{Q})$ of E. Namely, consider the finite set of all divisors y_0 of the discriminant

$$\Delta = 18a^2b^2 - 4a^3c + 18abc - 4b^3 - 27c^2, \quad a, b, c \in \mathbb{Z},$$

and find all integer solutions x_0 of the cubic equation

$$x^3 + ax^2 + bx + c - y_0^2 = 0.$$

Then all the nonzero points of finite order are contained among these.

In 1976 the American mathematician **B. Mazur** proved a wonderful theorem on the structure of torsion groups of elliptic curves [**C9**]:

5.5.5. THEOREM. *Let E be an elliptic curve defined over $\mathbb{Q}$. Then* $\operatorname{Tors} E(\mathbb{Q})$ *is isomorphic to one of the following fifteen groups: $\mathbb{Z}/m\mathbb{Z}$ for $m \leq 10$ or $m = 12$, and $\mathbb{Z}/2\mathbb{Z} \oplus \mathbb{Z}/2m\mathbb{Z}$ for $m \leq 4$.*

In the following table we give examples of rational elliptic curves for which these torsion groups are realized.

Table 3. Examples of torsion groups of elliptic curves E.

E	$\mathrm{Tors}\, E(Q)$
$y^2 = x^3 + 2$	0
$y^2 = x^3 + 8$	$\mathbb{Z}/2$
$y^2 = x^3 + 4$	$\mathbb{Z}/3$
$y^2 = x^3 + 4x$	$\mathbb{Z}/4$
$y^2 + y = x^3 - x^2$	$\mathbb{Z}/5$
$y^2 = x^3 + 1$	$\mathbb{Z}/6$
$y^2 - xy + 2y = x^3 + 2x^2$	$\mathbb{Z}/7$
$y^2 + 7xy - 6y = x^3 - 6x^2$	$\mathbb{Z}/8$
$y^2 + 3xy + 6y = x^3 + 6x^2$	$\mathbb{Z}/9$
$y^2 - 7xy - 36y = x^3 - 18x^2$	$\mathbb{Z}/10$
$y^2 + 43xy - 210y = x^3 - 210x^2$	$\mathbb{Z}/12$
$y^2 = x^3 - x$	$\mathbb{Z}/2 \oplus \mathbb{Z}/2$
$y^2 = x^3 + 5x^2 + 4x$	$\mathbb{Z}/2 \oplus \mathbb{Z}/4$
$y^2 - 5xy - 6y = x^3 - 3x^2$	$\mathbb{Z}/2 \oplus \mathbb{Z}/6$
$y^2 = x^3 + 337x^2 + 20736x$	$\mathbb{Z}/2 \oplus \mathbb{Z}/8$

CHAPTER 6

Algebraic Equations

The remaining part of the book is devoted to the solution of algebraic equations of fifth degree by means of theta functions. Originally this problem was solved by the famous French mathematician **Charles Hermite** in 1858. We divide the corresponding material into two chapters. Chapter 6 is devoted to common properties of algebraic equations, Lagrange's resolvents, Abel's theorem on unsolvability in radicals of the general equation of fifth degree, and Bring's form of the fifth degree equation. Regarding Lagrange's resolvents and roots of unity we follow [**B6**]. In Chapter 7, following the famous book by **Weber** [**B24**], we describe the scheme of solution of an arbitrary fifth degree equation (given in Bring's form) with the help of theta functions.

§6.1. Solving cubic and quartic equations

There are many known methods for solving third and fourth degree equations (also called *cubic* and *quartic* equations) in radicals. In this section we will only discuss the simplest of them. Certain other methods for solving cubic and quartic equations will be discussed in §6.3 and §6.6.

Let us first observe that the equation $x^n + a_1x^{n-1} + \dots + a_n = 0$ can be reduced to the form $y^n + b_2y^{n-2} + \dots + b_n = 0$ with the help of the change of variables $y = x + \frac{a_1}{n}$. Therefore, it suffices to consider cubic equations of the form $x^3 + ax + b = 0$ and quartic equations of the form $x^4 + ax^2 + bx + c = 0$.

Cubic equations. Let us try to represent the roots of the equation

$$x^3 + ax + b = 0 \tag{1.1}$$

in the form $x = \sqrt[3]{p} + \sqrt[3]{q}$. Then

$$x^3 = p + q + 3\sqrt[3]{pq}\,(\sqrt[3]{p} + \sqrt[3]{q}) = p + q + 3\sqrt[3]{pq}x.$$

Therefore, we should select p and q so that

$$\begin{cases} 3\sqrt[3]{pq} = -a \\ p + q = -b \end{cases} \quad \Longrightarrow \quad \begin{cases} pq = -\dfrac{a^3}{27} \\ p + q = -b. \end{cases}$$

Thus, we have derived a quadratic equation whose solutions are p and q. The roots of the quadratic equation are of the form

$$p, q = -\frac{b}{2} \pm \sqrt{\frac{b^2}{4} + \frac{a^3}{27}}. \tag{1.2}$$

The formula $x = \sqrt[3]{p} + \sqrt[3]{q}$ gives 9 distinct possibilities. To obtain the right three values we should make use of the relation $\sqrt[3]{p}\sqrt[3]{q} = -\frac{a}{3}$.

From this it follows that (1.1) has the solutions

$$x = \sqrt[3]{p} - \frac{a}{3\sqrt[3]{p}}, \tag{1.3}$$

where p is found by formula (1.2). Here the value of x does not depend on the choice of the sign in front of the radical in (1.2).

(It is easy to verify that the values of x determined by formula (1.3) are actually the roots of equation (1.1).)

Quartic equations. *First method.* Let us try to represent the polynomial $x^4 + ax^2 + bx + c$ as the difference of two squares. To this end, let us use the identity

$$x^4 + ax^2 + bx + c = \left(x^2 + \frac{a}{2} + t\right)^2 - \left(2tx^2 - bx + \left(t^2 + at - c + \frac{a^2}{2}\right)\right).$$

Select t for which the discriminant $D = b^2 - 8t\left(t^2 + at - c + \frac{a^2}{2}\right)$ vanishes. Then

$$x^4 + ax^2 + bx + c = \left(x^2 + \frac{a}{2} + t\right)^2 - 2t\left(x - \frac{b}{4t}\right)^2.$$

Therefore, the equation

$$x^4 + ax^2 + bx + c = 0 \tag{1.4}$$

can be solved as follows. First, solve the cubic equation for t

$$b^2 - 8t\left(t^2 + at - c + \frac{a^2}{2}\right) = 0.$$

Let t_0 be one of its roots. Then equation (1.4) can be expressed in the form

$$x^2 + \frac{a}{2} + t_0 = \pm\sqrt{2t_0}\left(x - \frac{b}{4t_0}\right).$$

Second method. (Euler) Let x_1, x_2, x_3, x_4 be the roots of equation (1.4). Set $u = x_1 + x_2 = -(x_3 + x_4)$. Then

$$x^4 + ax^2 + bx + c = \left(x^2 - ux + \alpha\right)\left(x^2 + ux + \beta\right),$$

i.e.,

$$\alpha + \beta - u^2 = a, \quad u(\alpha - \beta) = b, \quad \alpha\beta = c.$$

From the first and the second of the above equations we get

$$\alpha = \frac{1}{2}\left(a + u^2 + \frac{b}{u}\right), \quad \beta = \frac{1}{2}\left(a + u^2 - \frac{b}{u}\right).$$

Substituting these expressions into the third equation we get

$$u^6 + 2au^4 + \left(a^2 - 4c\right)u^2 - b^2 = 0. \tag{1.5}$$

Equation (1.5) is a cubic equation for u^2. First, we solve this cubic equation; then we find the 6 roots of equation (1.5). They are of the form $\pm u_1$, $\pm u_2$, $\pm u_3$. We may assume that

$$\begin{aligned} x_1 + x_2 &= u_1, & x_3 + x_4 &= -u_1, \\ x_1 + x_3 &= u_2, & x_2 + x_4 &= -u_2, \\ x_1 + x_4 &= u_3, & x_2 + x_3 &= -u_3. \end{aligned}$$

Then $u_1 + u_2 + u_3 = 2x_1$.

Third method. Instead of equation (1.4) we may solve the system of equations

$$\begin{cases} f = y - x^2 = 0, \\ g = y^2 + ay + bx + c = 0. \end{cases}$$

The second equation of this system can be replaced by $\lambda f + g = 0$. The second degree curve $\lambda f + g = 0$ represents a pair of straight lines if and only if

$$(1.6) \qquad \begin{vmatrix} -\lambda & 0 & \frac{b}{2} \\ 0 & 1 & \frac{a+\lambda}{2} \\ \frac{b}{2} & \frac{a+\lambda}{2} & c \end{vmatrix} = 0.$$

If λ_0 is a root of the cubic equation (1.6), the equation $\lambda_0 f + g = 0$ factorizes into two linear equations and, therefore, the system

$$\begin{cases} f = 0 \\ \lambda_0 f + g = 0 \end{cases}$$

is easy to solve.

§6.2. Symmetric polynomials

A polynomial $f(x_1, \dots, x_n)$ is called *symmetric* if for any permutation $(i_1, \dots, i_n)$ we have $f(x_{i_1}, \dots, x_{i_n}) = f(x_1, \dots, x_n)$. The *elementary symmetric polynomials* $\sigma_i(x_1, \dots, x_n)$, defined by the relation

$$(x - x_1) \cdots (x - x_n) = x^n - \sigma_1 x^{n-1} + \sigma_2 x^{n-2} - \cdots + (-1)^n \sigma_n,$$

are the most important examples of symmetric polynomials. Thus if $x_1, \dots, x_n$ are the roots of the polynomial $x^n + a_1 x^{n-1} + \cdots + a_n$, then $\sigma_i(x_1, \dots, x_n) = (-1)^i a_i$. It is convenient to set $\sigma_k(x_1, \dots, x_n) = 0$ for $k > n$.

6.2.1. THEOREM (Main theorem on symmetric polynomials). *Any symmetric polynomial $f(x_1, \dots, x_n)$ can be represented in the form of a polynomial in elementary symmetric polynomials, i.e., $f(x_1, \dots, x_n) = g(\sigma_1, \dots, \sigma_n)$, where g is a uniquely defined polynomial.*

PROOF. It suffices to consider the case when f is a homogeneous polynomial. Let us order monomials *lexicographically*, i.e., say that $ax_1^{\alpha_1} \cdots x_n^{\alpha_n}$ is *greater than* $bx_1^{\beta_1} \cdots x_n^{\beta_n}$ if $\alpha_1 = \beta_1, \dots, \alpha_k = \beta_k$, $\alpha_{k+1} > \beta_{k+1}$ (k can be equal to 0). The greatest monomial of a polynomial is called its *leading term.* The following properties of lexicographic order are easy to verify:

(1) The leading term of the product of two polynomials is equal to the product of their leading terms.

(2) If $ax_1^{\alpha_1} \cdots x_n^{\alpha_n}$ is the leading term of a symmetric polynomial, then $\alpha_1 \geq \alpha_2 \geq \cdots \geq \alpha_n$.

Let the leading term of a homogeneous symmetric polynomial f be equal to $ax_1^{\alpha_1}\cdots x_n^{\alpha_n}$. Let us consider the polynomial $a\sigma_1^{\alpha_1-\alpha_2}\sigma_2^{\alpha_2-\alpha_3}\cdots\sigma_n^{\alpha_n}$. By property (1) its leading term is equal to

$$ax_1^{\alpha_1-\alpha_2}(x_1^{\alpha_2-\alpha_3}x_2^{\alpha_2-\alpha_3})\cdots(x_1^{\alpha_n}\cdots x_n^{\alpha_n}) = ax_1^{\alpha_1}\cdots x_n^{\alpha_n}.$$

Therefore, the degree of the leading term of the polynomial

$$f_1 = f - a\sigma_1^{\alpha_1-\alpha_2}\sigma_2^{\alpha_2-\alpha_3}\cdots\sigma_n^{\alpha_n}$$

is lower than that of f. Let us apply to f_1 the same operation as to f, and so on. Since the number of monomials whose degree is lower than that of the leading term of f is finite, it follows that after finitely many operations we get the zero polynomial. This means that

$$f = a\sigma_1^{\alpha_1-\alpha_2}\sigma_2^{\alpha_2-\alpha_3}\cdots\sigma_n^{\alpha_n} + \cdots = g(\sigma_1,\dots,\sigma_n).$$

Now we prove the uniqueness of the representation $f = g(\sigma_1,\dots,\sigma_n)$. It suffices to verify that if $g(\sigma_1,\dots,\sigma_n) \neq 0$ for any collection $x_1,\dots,x_n$, then g is the zero polynomial. Let g be a nonzero polynomial (over a field of zero characteristic). Then $g(a_1,\dots,a_n) \neq 0$ for a collection $\{a_1,\dots,a_n\}$. It is also clear that if $x_1,\dots,x_n$ are the roots of the polynomial

$$x^n - a_1x^{n-1} + a_2x^{n-2} - \cdots + (-1)^n a_n,$$

then $\sigma_i(x_1,\dots,x_n) = a_i$. □

Remark. In Theorem 6.2.1, if the coefficients of f are integers, then the coefficients of g are also integers.

The main theorem on symmetric polynomials plays an important role in the theory of algebraic equations for the following reason. Let $x_1,\dots,x_n$ be the roots of the polynomial $x^n + a_1x^{n-1} + \cdots + a_n$. Suppose a polynomial $f(x_1,\dots,x_n)$ does not change under any permutations of its roots. Then it can be represented as a polynomial of the coefficients $a_1,\dots,a_n$. We will often use this property.

Problems

6.2.1. Let $s_k(x_1,\dots,x_n) = x_1^k + \cdots + x_n^k$. Prove that

$$s_k - s_{k-1}\sigma_1 + s_{k-2}\sigma_2 - \cdots + (-1)^k k\sigma_k = 0.$$

§6.3. The Lagrange resolvents

We can solve the quadratic equation $x^2 + ax + b = 0$ as follows. Let x_1 and x_2 be the roots of this equation. Then

$$(3.1)\qquad x_1 = \frac{1}{2}\left[(x_1+x_2)+(x_1-x_2)\right] = \frac{1}{2}\left[(x_1+x_2)+\sqrt{(x_1-x_2)^2}\right].$$

Here $x_1 + x_2$ and $(x_1 - x_2)^2$ are symmetric functions of the roots and, therefore, they can be expressed in terms of the coefficients a and b. The explicit expressions are as follows:

$$x_1 + x_2 = -a, \qquad (x_1 - x_2)^2 = (x_1+x_2)^2 - 4x_1x_2 = a^2 - 4b.$$

Thus,

$$x_1 = \frac{1}{2}\left[-a + \sqrt{a^2 - 4b}\,\right].$$

The root $\sqrt{a^2 - 4b}$ has two values. One of them yields one root of the equation, the other value yields the other root.

We can apply a similar approach to the solution of a cubic equation. Here, instead of square roots, we get cubic roots. Let α be a primitive cubic root of unity, i.e., $\alpha^3 = 1$ and $\alpha \neq 1$.

For the roots of the cubic equation $x^3 + ax^2 + bx + c = 0$ the analog of formula (3.1) looks as follows:

$$\begin{aligned} x_1 &= \frac{1}{3}\left[(x_1 + x_2 + x_3) + (x_1 + \alpha x_2 + \alpha^2 x_3) + (x_1 + \alpha^2 x_2 + \alpha x_3)\right] \\ &= \frac{1}{3}\left[(x_1 + x_2 + x_3) + \sqrt[3]{(x_1 + \alpha x_2 + \alpha^2 x_3)^3} + \sqrt[3]{(x_1 + \alpha^2 x_2 + \alpha x_3)^3}\right]. \end{aligned} \tag{3.2}$$

To make use of this formula we have to calculate

$$u = (x_1 + \alpha x_2 + \alpha^2 x_3)^3$$

and

$$v = (x_1 + \alpha^2 x_2 + \alpha x_3)^3.$$

Under any permutation of the roots, u turns either into itself or into v. (To verify this, factor out α^k from x_1 in the expression obtained after permutation of roots.) Thus, $u + v$ and uv are symmetric polynomials of x_1, x_2 and x_3 and, therefore, they can be expressed in terms of the coefficients of the cubic equation. After these expressions are obtained, solving the cubic equation reduces to solving a quadratic equation.

This method for solving a cubic equation was independently suggested by two French mathematicians, **Lagrange** and **Vandermonde**. They offered it simultaneously in 1770. Vandermonde presented his work to the Paris Academy, while Lagrange, who worked during 1766–1787 in Berlin, reported his results to the Berlin Academy.

Vandermonde's work *Mémoire sur la résolution des équations* contained a new approach to the solution of the third and fourth degree equations and also to certain other equations, in particular, the equation $x^{11} - 1 = 0$. Vandermonde's paper was not published until 1774. During this time, in 1771 and 1773 two parts of Lagrange's fundamental treatise *Réflexions sur la résolution des équations* were published. This treatise contained almost the same ideas as Vandermonde's but these ideas were considerably more elaborate. Having acquainted himself with the works of Lagrange, Vandermonde never returned to this topic.

On the base of the example above, introduce the following definition. Let $x^n + a_1 x^{n-1} + \cdots + a_n$ be a polynomial with rational coefficients, $x_0, \ldots, x_{n-1}$ its roots. The *Lagrange resolvents* are expressions

$$r(x_0, \alpha) = x_0 + \alpha x_1 + \cdots + \alpha^{n-1} x_{n-1},$$

where α is the nth root of unity, i.e., $\alpha^n = 1$.

The points α satisfying $\alpha^n = 1$ form the vertices of a regular n-gon and the points α^k form the vertices of a regular m-gon, where $m = n/(n,k)$. Therefore,

$$\sum_\alpha \alpha^k = \begin{cases} n & k \equiv 0 \pmod{n}, \\ 0 & k \not\equiv 0 \pmod{n}. \end{cases}$$

Hence,

$$(3.3)\qquad \begin{aligned} nx_0 &= \sum_\alpha r(x_0, \alpha), \\ nx_k &= \sum_\alpha \alpha^{-k} r(x_0, \alpha). \end{aligned}$$

Thus, if the Lagrange resolvents are known, the roots of the equation can be calculated by formulas (3.3).

Lagrange suggested a method for calculating resolvents that we will illustrate with the cubic equation. Let $r = x_0 + \alpha x_1 + \alpha^2 x_2$ be a Lagrange resolvent for the cubic polynomial. Six possible permutations of the roots x_0, x_1, x_2 give us six values $r_1, \dots, r_6$. Let us consider the sixth degree polynomial

$$g(t) = (t - r_1)(t - r_2) \cdots (t - r_6).$$

The coefficients of g are symmetric polynomials of r_i and, therefore, they are symmetric polynomials in x_0, x_1 and x_3; hence, they can be expressed in terms of the coefficients of the initial equation.

Lagrange called the equation $g(t) = 0$ the *resolving equation*. The point is that this equation is easy to be solved in radicals. Indeed, let $r_1 = x_0 + \alpha x_1 + \alpha^2 x_2$ and $r_4 = x_0 + \alpha x_2 + \alpha^2 x_1$. Then the remaining r_i are equal to

$$r_2 = \alpha r_1, \quad r_3 = \alpha^2 r_1, \quad r_5 = \alpha r_4, \quad r_6 = \alpha^2 r_4.$$

Hence,

$$\begin{aligned} (t - r_1)(t - r_2)(t - r_3) &= t^3 - r_1^3, \\ (t - r_4)(t - r_5)(t - r_6) &= t^3 - r_4^3, \end{aligned}$$

and, therefore, $g(t) = \left(t^3 - r_1^3\right)\left(t^3 - r_4^3\right)$. Thus, the resolving equation $g(t) = 0$ is a quadratic equation for t^3.

Now let us find the resolving equation for the fourth degree polynomial. Here for the fourth degree root of unity it is convenient to take $\alpha = -1$ instead of the primitive root $\alpha = \pm i$. Then the resolvent $r = x_0 - x_1 + x_2 - x_3$ takes, under permutations, only $24/4 = 6$ rather than $4! = 24$ distinct values and each value is taken exactly 4 times. These values are as follows:

$$\begin{aligned} r_1 &= (x_0 + x_1) - (x_2 + x_3) = -r_4, \\ r_2 &= (x_0 + x_2) - (x_1 + x_3) = -r_5, \\ r_3 &= (x_0 + x_3) - (x_1 + x_2) = -r_6. \end{aligned}$$

Thus, $g(t) = h^4(t)$, where $h(t) = \left(t^2 - r_1^2\right)\left(t^2 - r_2^2\right)\left(t^2 - r_3^2\right)$. Hence, the equation $g(t) = 0$ is equivalent to the equation $h(t) = 0$, which is a cubic equation for t^3.

There is no necessity to give the calculations of the resolvent for the primitive roots $\alpha = \pm i$. Indeed,

$$(x_0 + x_1 + x_2 + x_3) + r_1 + r_2 + r_3 = 4x_0.$$

For the 5th degree polynomial the resolving equation is of degree $5! = 120$. Let

$$r = x_0 + \alpha x_1 + \alpha^2 x_2 + \alpha^3 x_3 + \alpha^4 x_4,$$

where $\alpha^5 = 1$. If r_1 is one of the values of r, then αr_1, $\alpha^2 r_1$, $\alpha^3 r_1$ and $\alpha^4 r_1$ are also values of r. Therefore, the resolving polynomial $g(t)$ can be factorized and

$$h(t) = (t - r_1)(t - \alpha r_1) \cdots (t - \alpha^4 r_1) = t^5 - r_1^5$$

is one of the factors. It is easy to see that $g(t)$ is the product of 24 factors of this form. Hence, $g(t)$ is a polynomial of degree 24 in t^5.

In spite of considerable efforts, Lagrange did not succeed in solving the equation $g(t) = 0$. This failure convinced him that it is impossible to solve a generic quintic equation in radicals.

Nevertheless, for the equations that can be solved in radicals Lagrange's method of resolvents turned out to be an important method for their solution. In the next section we will use it to solve in radicals the equation $x^n - 1 = 0$.

§6.4. Roots of unity

The equation $x^n - 1 = 0$ admits an obvious solution in radicals: $x = \sqrt[n]{1}$. This solution is, however, unsatisfactory for the following reason. Dividing $x^n - 1$ by $x - 1$ we get the polynomial

$$f_n(x) = x^{n-1} + x^{n-2} + \cdots + x + 1.$$

The degree of f_n is equal to $n - 1$ and, therefore, the formula $x = \sqrt[n]{1}$ provides us with n roots, one root more than we would like to have. We would like to get a formula for the roots of f_n which does not involve roots of degree higher than $n-1$. This will be obtained in this section.

First, let us consider several examples. The formulas for the roots of the polynomials $f_3(x) = x^2 + x + 1$ and $f_4(x) = (x+1)(x^2+1)$ are elementary to obtain. So let us immediately pass to the polynomial $f_5(x) = x^4 + x^3 + x^2 + x + 1$. Set $y = x + x^{-1}$. Then the equation $f_5(x) = 0$ is equivalent to the equation $y^2 + y - 1 = 0$. The roots of the latter equation are

$$y_{1,2} = \frac{-1 \pm \sqrt{5}}{2},$$

and the roots of equations $y_i = x + x^{-1}$ are

$$\frac{\sqrt{5} - 1 \pm \sqrt{-2\sqrt{5} - 10}}{4}, \quad \frac{-\sqrt{5} - 1 \pm \sqrt{2\sqrt{5} - 10}}{4}.$$

The number α is called a *primitive* nth root (or a root of degree n) of unity if $\alpha^n = 1$ and $\alpha^k \neq 1$ for $k = 1, 2, \ldots, n-1$. In this case the numbers $\alpha, \alpha^2, \ldots, \alpha^{n-1}$ are distinct. It is easy to verify that if α and β are, respectively, pth and qth primitive roots of unity, where p and q are relatively prime, then $\alpha\beta$ is a primitive pqth root of unity. Therefore, it suffices to consider equations $f_m(x) = 0$, where m is a power of a prime.

For the equation $f_7(x) = 0$ we can, as before, make the change of variables $y = x + x^{-1}$. As a result we get the equation

$$y^3 + y^2 - 2y - 1 = 0.$$

Its roots are given by the formula

$$y_{1,2,3} = \frac{1}{3}\left(-1 + \sqrt[3]{\frac{-7+21\sqrt{-3}}{6}} + \sqrt[3]{\frac{-7-21\sqrt{-3}}{6}}\right).$$

Now the roots of the equation $f_7(x) = 0$ are obtained as solutions of the quadratic equations $x^2 - y_i x + 1 = 0$.

For the polynomial $f_{11}(x)$ the change of variables $y = x + x^{-1}$ does not give the desired result since this change of variables leads to a quintic equation, which is unclear how to solve. Lagrange's method of resolvents enables one to overcome these difficulties. The numbers $\alpha, \alpha^2, \dots, \alpha^{10}$, where α is one of the primitive 11th roots of unity, are the roots of equation

$$f_{11}(x) = x^{10} + x^9 + \dots + x + 1 = 0. \tag{4.1}$$

The degree of the polynomial f_{11} is equal to 10, hence, to construct its Lagrange resolvents, we need a 10th root of unity. Let β be a primitive 10th root of unity. Let us order the roots of equation (4.1) so that each next root is the square of the preceding one:

$$\alpha,\ \alpha^2,\ \alpha^4,\ \alpha^8,\ \alpha^5,\ \alpha^{10},\ \alpha^9,\ \alpha^7,\ \alpha^3,\ \alpha^6. \tag{4.2}$$

Consider the Lagrange resolvent constructed from this sequence of roots:

$$r_1 = r(\alpha, \beta) = \alpha + \beta\alpha^2 + \beta^2\alpha^4 + \beta^3\alpha^8 + \dots + \beta^8\alpha^3 + \beta^9\alpha^6.$$

Under the change of the primitive root β with the root β^i, where $i = 1, \dots, 10$, we get the Lagrange resolvents $r_i = r(\alpha, \beta^i)$, where $i = 1, \dots, 10$. It is easy to verify that

$$r_1 + r_2 + \dots + r_{10} = 10\alpha.$$

Therefore, the 11th root of unity, α, will be found if we succeed in calculating $r_1, \dots, r_{10}$.

First, we calculate the r_i^{10}. Since $\alpha^{11} = 1$ and $\beta^{10} = 1$, it follows that r_i can be represented in the form $\sum_{j,k} a_{ijk}\beta^j\alpha^k$, where $0 \le j \le 9$, $0 \le k \le 10$, and r_i^{10} can be represented in the form $\sum_{j,k} b_{ijk}\beta^j\alpha^k$, where $0 \le j \le 9$, $0 \le k \le 10$.

Here a_{ijk} and b_{ijk} are nonnegative integers. Let us express r_i^{10} as a polynomial of α, where the degrees of α are ordered as in (4.2):

$$r_i^{10} = p_{i,0}(\beta) + p_{i,1}(\beta)\alpha + p_{i,2}(\beta)\alpha^2 + p_{i,3}(\beta)\alpha^4 + \dots + p_{i,10}(\beta)\alpha^6. \tag{4.3}$$

Let us prove that r_i^{10} does not depend on α, i.e., $r_i^{10} = p_i(\beta)$, where p_i is a polynomial with integer coefficients. It is easy to verify that $r(\alpha^2, \beta^i) = \beta^{-i}r(\alpha, \beta^i)$, i.e., $r_i(\alpha^2) = \beta^{-i}r_i(\alpha)$. Hence, $r_i^{10}(\alpha^2) = r_i^{10}(\alpha)$. Thus, denoting for brevity $p_{i,k}(\beta)$ in formula (4.3) by p_k we get

$$\begin{aligned} &p_0 + p_1\alpha + p_2\alpha^2 + p_3\alpha^4 + p_4\alpha^8 + \dots + p_9\alpha^3 + p_{10}\alpha^6 \\ &\quad = p_0 + p_1\alpha^2 + p_2\alpha^4 + p_3\alpha^8 + p_4\alpha^5 + \dots + p_9\alpha^6 + p_{10}\alpha. \end{aligned}$$

Therefore,

$$(p_{10} - p_1)\,\alpha + (p_1 - p_2)\,\alpha^2 + (p_2 - p_3)\,\alpha^4 + \dots + (p_9 - p_{10})\,\alpha^6 = 0. \tag{4.4}$$

The following statement holds.

6.4.1. THEOREM. *Let p be a prime, and let α and β be primitive roots of unity whose degrees are p and $p-1$, respectively. Then the equation*

$$q_1(\beta)\alpha + q_2(\beta)\alpha^2 + \cdots + q_{p-1}(\beta)\alpha^{p-1} = 0,$$

where $q_1, \ldots, q_{p-1}$ are polynomials with integer coefficients, implies that $q_1(\beta) = q_2(\beta) = \cdots = q_{p-1}(\beta) = 0$.

Although this statement (there are no nontrivial relations between α and β, something like linear independence) seems almost obvious, its proof is rather complicated. It requires elements of Galois theory and, therefore, we will omit it, see [**B6**], p.97.

Applying Theorem 6.4.1 for $p = 11$ to equation (4.4), we get $p_{10} = p_1 = p_2 = p_3 = \cdots = p_9$. Therefore,

$$r_i^{10} = p_0(\beta) + p_1(\beta)\left(\alpha + \alpha^2 + \alpha^3 + \cdots + \alpha^{10}\right) = p_0(\beta) - p_1(\beta).$$

Let us summarize. It is possible to represent a primitive 10th root of unity in the form of the product of primitive roots of degrees 5 and 2. Therefore, β can be expressed with the help of quadratic radicals. Next, we can represent r_i^{10} as a polynomial in β with integer coefficients. It remains to use the formula

$$\alpha = \frac{1}{10}\left(\sqrt[10]{r_1^{10}} + \cdots + \sqrt[10]{r_{10}^{10}}\right).$$

This formula gives 100 values instead of 10, as we would have liked. We have to select the 10 values needed among all the values obtained. Therefore, the formula

$$\alpha = \frac{1}{10}(r_1 + \cdots + r_{10}) = \frac{1}{10}\left(r_1 + \frac{r_2 r_1^8}{r_1^8} + \frac{r_3 r_1^7}{r_1^7} + \cdots\right) \tag{4.5}$$

is more convenient. The point is that the values $r_i r_1^{10-i}$, as well as r_i^{10}, do not depend on α. To prove this, it suffices to observe that

$$r_i(\alpha^2)r_1^{10-i}(\alpha^2) = \beta^{-i} r_i(\alpha)\left(\beta^{-1} r_1(\alpha)\right)^{10-i} = r_i(\alpha)r_1^{10-i}(\alpha).$$

Therefore, formula (4.5) determines α uniquely after any one of 10 values of $r_1 = \sqrt[10]{r_1^{10}}$ is chosen.

We can similarly solve the equation $f_p(x) = 0$ for any prime $p > 2$ as well. Resolving the equation $f_{11}(x) = 0$ we made use of the fact that the primitive 11th roots of unity can be ordered so that each next root is the square of the preceding one. Such an ordering of roots is possible because the numbers 2^k, where $k = 0, 1, \ldots, 9$, have distinct residues modulo 11. A similar statement for residues modulo, for example, 7 is false but for any prime p there exists a number g, called a *primitive element*, such that the residues of the numbers g^k, where $k = 0, 1, \ldots, p-2$, modulo p are distinct. In other words, the multiplicative group of nonzero residues modulo p is the cyclic group generated by g. We will not prove this well-known statement.

Let α and β be primitive roots of unity of degrees p and $p-1$, respectively. If the residues modulo p of the numbers g^k for $k = 0, 1, \ldots, p-2$ are distinct, then the numbers

$$\alpha, \alpha^g, \alpha^{g^2}, \ldots, \alpha^{g^{p-1}}$$

exhaust all primitive pth roots of unity. Therefore, for a polynomial $f_p(x)$ we can consider the Lagrange resolvent

$$r_1 = \alpha + \beta\alpha^g + \beta^2\alpha^{g^2} + \cdots + \beta^{p-1}\alpha^{g^{p-1}}.$$

Set $r_i = r_1(\alpha, \beta^i)$. It is easy to verify that $r_1(\alpha^g, \beta^i) = \beta^{-i} r_1(\alpha, \beta^i)$. Hence, the quantities r_1^{p-1} and $r_i r_1^{p-1-i}$ do not vary when α is replaced with α^g. As in the case $p = 11$, we can prove with the help of this property and Theorem 6.4.1 that r_1^{p-1} and $r_i r_1^{p-1-i}$ are polynomials in β with integer coefficients. For β an expression in radicals can be obtained by the induction because β is a root of unity of degree $p - 1 < p$. The formula for α is as follows:

$$\alpha = \frac{1}{p-1}\left(r_1 + \frac{r_2 r_1^{p-3}}{r_1^{p-3}} + \frac{r_3 r_1^{p-4}}{r_1^{p-4}} + \cdots\right).$$

This formula gives an unambiguous expression for α after one of the values of $r_1 = \sqrt[p-1]{r_1^{p-1}}$ is chosen.

§6.5. The Abel theorem on the unsolvability in radicals of the general quintic equation

Lagrange's works spurred many geometers (this was the common name for all mathematicians of that time) to begin searching for a proof of the impossibility to solve in radicals the general quintic equation and higher degree equations. In 1788–1813 there appeared several papers of the Italian mathematician **Paulo Ruffini** (1765–1822). Following Lagrange, he considered substitutions of roots of equations and it was he who coined the term the *group of substitutions*. His series of papers culminated in the proof of the theorem on impossibility to solve in radicals the general equations of the fifth and higher degrees.

Regrettably, this proof had an essential gap. Without justification Ruffini assumed that the radicals can be rationally expressed in terms of the roots of the initial equation (cf. Theorem 6.5.4 below).

The Norwegian mathematical genius **Niels Henrik Abel** (1802–1829) was the first to give a complete proof of the theorem on unsolvability of the general quintic equation. He exposed his proof in the memoir *Proof on Impossibility of an Algebraic Solution of General Fifth Degree Equations* published in the first issue of Crelle's journal in 1826.

We say that the equation

$$F(x) = x^n + c_1 x^{n-1} + \cdots + c_n = 0 \tag{5.1}$$

is the *general nth degree equation* if its coefficients $c_1, \dots, c_n$ are independent variables over the ground field L. In what follows we will assume that $L = \mathbb{Q}$.

Adjoining $c_1, \dots, c_n$ to $\mathbb{Q}$, we get the field $\Delta = \mathbb{Q}(c_1, \dots, c_n)$. This field is called the *rationality field of equation* (5.1).

Having attached to Δ the roots $\alpha_1, \dots, \alpha_n$ of equation (5.1) we get the field $\Delta(F) = \Delta(\alpha_1, \dots, \alpha_n)$, called the *normal field* of equation (5.1) or the *Galois field* of this equation.

We will say that equation (5.1) is *solvable in radicals* if $\Delta(F)$ is contained in the extension R of Δ obtained after attaching to Δ certain radicals

$$\rho_1 = \sqrt[p]{a_1},\ \rho_2 = \sqrt[q]{a_2},\ \dots,\ \rho_m = \sqrt[s]{a_m},$$

where

$$a_1 \in \Delta,\ a_2 \in \Delta(\rho_1),\ a_3 \in \Delta(\rho_1, \rho_2),\ \ldots,\ a_m \in \Delta(\rho_1, \ldots, \rho_{m-1}).$$

EXAMPLE. Let $F(x) = x^2 + c_1x + c_2$. Then $\Delta = \mathbb{Q}(c_1, c_2)$ and $\Delta(F) = \Delta(\sqrt{a_1})$, where $a_1 = c_1^2 - 4c_2 \in \Delta$.

Observe that the exponents $p, q, \ldots, s$ of the radicals $\rho_1, \rho_2, \ldots, \rho_m$ can be assumed to be primes. Indeed, if $p = lm$, then instead of adjoining the radical $\rho_1 = \sqrt[p]{a_1}$ we may consecutively adjoin the radicals $\rho = \sqrt[l]{a_1}$ and $\rho_1 = \sqrt[m]{\rho}$. Therefore, in what follows we will only consider adjoining the radicals with prime exponents.

Suppose that equation (5.1) is solvable in radicals. Adjoin to Δ the primitive roots of unity $\varepsilon_1, \ldots, \varepsilon_m$ whose degrees are equal to the degrees of the radicals $\rho_1, \ldots, \rho_m$, respectively. Denote the obtained field by K.

Since $\Delta \subset K$, it follows that

$$\Delta(F) \subset \Delta(\rho_1, \ldots, \rho_m) \subset K(\rho_1, \ldots, \rho_m).$$

To prove Abel's theorem, we will need four auxiliary statements, Theorems 6.5.1–6.5.4.

6.5.1. THEOREM. *Let p be a prime and k a field of zero characteristic. The polynomial $x^p - a$ is reducible over k if and only if $a = b^p$ for some $b \in k$.*

PROOF. Suppose $x^p - a = f(x)g(x)$, where $f(x)$ and $g(x)$ are polynomials over k. Let ε be a primitive pth root of unity and $\beta = \sqrt[p]{a}$. Then

$$f(x) = x^r + c_1x^{r-1} + \cdots + c_r = (x - \varepsilon^{n_1}\beta)\cdots(x - \varepsilon^{n_r}\beta).$$

Hence, $\pm\varepsilon^l\beta^r = c_r \in k$, where $l = n_1 + \cdots + n_r$. Since $(\varepsilon^l)^p = 1$, it follows that $(\pm\beta^r)^p = (c_r)^p$, i.e., $a^r = (\pm c_r)^p$. The number p is prime and $1 \leq r = \deg f < p$; hence, $rs + pt = 1$ for certain integers s and t. Therefore, $a = a^{rs}a^{pt} = (\pm c_r a^t)^p = b^p$, where $b = \pm c_r a^t \in k$.

It is also clear that if $a = b^p$, then $x^p - a$ is reducible because it is divisible by $x - b$. □

6.5.2. THEOREM. *Let s be a prime and $a_i \in k = K(\rho_1, \ldots, \rho_{i-1})$. If $\rho_i = \sqrt[s]{a_i} \notin k$, then $\rho_i^l \in k$ if and only if l is divisible by s.*

PROOF. If $l = ns$, then $\rho_i^l = a_i^n \in k$ since $a_i \in k$. Now suppose that $\rho_i^l = a \in k$ and $l = sq + r$, where $0 < r < s$. Then $a = \rho_i^l = (a_i)^q\rho_i^r$ and, therefore, $\rho_i^r = b$, where $b = a\,(a_i)^{-q} \in k$.

Over k, the polynomials $x^s - a_i$ and $x^r - b$ have a common root ρ_i; hence, they have a common divisor whose degree does not exceed $r < s$. In particular, the polynomial $x^s - a_i$ is reducible over k. Theorem 6.5.1 implies that $a_i = b^s$, where $b \in k$. Clearly, $b = \varepsilon\rho_i$, where ε is a primitive root of unity of degree s. Since $\varepsilon \in K \subset k$, it follows that $\rho_i \in k$. Contradiction. □

We may assume that $\rho_1, \ldots, \rho_m$ is a *minimal* sequence of radicals (of prime degrees) required to compute a root α of equation (5.1), i.e., any other such sequence contains at least m radicals. In what follows we will only consider minimal sequences of radicals. Under this assumption the following statement holds.

6.5.3. THEOREM. *Let $\rho_1, \dots, \rho_m$ be a minimal sequence of radicals needed to compute a root α of equation* (5.1). *Then α can be represented in the form*

$$\alpha = u_0 + \rho + u_2\rho^2 + \dots + u_{s-1}\rho^{s-1},$$

where s is the degree of $\rho_m, \rho = \sqrt[s]{a}, a \in k = K(\rho_1, \dots, \rho_{m-1})$ and $u_i \in k$.

PROOF. Since $\alpha \in K(\rho_1, \dots, \rho_m) = k(\rho_m)$ and $\rho_m^s \in k$, we have

$$\alpha = b_0 + b_1\rho_m + b_2\rho_m^2 + \dots + b_{s-1}\rho_m^{s-1}, \tag{5.2}$$

where $b_i \in k$. The only difficulty is to ensure that $b_1 = 1$. By the assumption, $\alpha \notin k$ so that at least one of the numbers $b_1, \dots, b_{s-1}$ is nonzero. Let $b_l \neq 0$ for some l such that $1 \leq l < s$. Set $\rho = b_l\rho_m^l$. Since s is a prime, $ul + vs = 1$ for certain integers u and v. Moreover, we have

$$\rho^u = b_l^u\rho_m^{ul} = b_l^u\rho_m^{1-vs} = b_l^u a^{-v}\rho_m,$$

i.e., $\rho_m = c\rho^u$, where $c = b_l^{-u}a^v \in k$. Since $\rho_m \notin k$, it follows that $\rho \notin k$. It is also clear that $\rho^s = b_l^s\rho_m^{ls} = b_l^s a^l \in k$.

In (5.2) replace ρ_m with $c\rho^u$ taking into account that $b_l\rho_m^l = \rho$. As a result, we get

$$\alpha = b_0 + b_1c\rho^u + b_2c^2\rho^{2u} + \dots + \rho + \dots + b_{s-1}c^{s-1}\rho^{(s-1)u}. \tag{5.3}$$

Theorem 6.5.2 implies that $\rho^t \in k$ if and only if t is divisible by s. Since u and s are relatively prime, the elements $1, \rho^u, \rho^{2u}, \dots, \rho^{(s-1)u}$ are linearly independent over k and the set of these elements coincides with the set $1, \rho, \rho^2, \dots, \rho^{s-1}$ (perhaps, ordered differently). Thus, formula (5.3) gives the required expression for α:

$$\alpha = b_0 + \rho + b_2'\rho^2 + \dots + b_{s-1}'\rho^{s-1}. \qquad \square$$

6.5.4. THEOREM. *The minimal sequence of radicals $\rho_1, \dots, \rho_m$ necessary to calculate a root α of polynomial* (5.1) *can be selected so that $\rho_1, \dots, \rho_m$ are polynomials over K of the roots $\alpha_1, \dots, \alpha_n$ of polynomial* (5.1).

PROOF. Start with an arbitrary minimal sequence $\rho_1, \dots, \rho_m$. By Theorem 6.5.3 we can replace ρ_m with a radical ρ of the same degree s so that

$$\alpha = u_0 + \rho + u_1\rho^2 + \dots + u_{s-1}\rho^{s-1},$$

where $u_i \in k = K(\rho_1, \dots, \rho_{m-1})$ and $\rho^s = a \in k$. Let us show that for any root ξ of the polynomial $x^s - a$

$$\alpha(\xi) = u_0 + \xi + u_1\xi^2 + \dots + u_{s-1}\xi^{s-1}$$

is a root of polynomial (5.1). Substitute $x = \alpha(\xi)$ in the polynomial

$$F(x) = x^n + c_1x^{n-1} + \dots + c_n.$$

Taking into account that $\xi^s = a \in k$ we get an expression of the form

$$b_0 + b_1\xi + \dots + b_{s-1}\xi^{s-1},$$

where $b_i \in k$. The polynomials $x^s - a$ and $b_0 + b_1x + \dots + b_{s-1}x^{s-1}$ have a common root ρ; hence, they have a common divisor over k. By Theorem 6.5.1 the polynomial $x^s - a$ is irreducible over k; hence, $b_0 = b_1 = \dots = b_{s-1} = 0$. This means that if ξ

is a root of the polynomial $x^s - a$, then $\alpha(\xi)$ is a root of polynomial (5.1). Let ε be a primitive root of unity of degree s. Then $\xi = \varepsilon^r \rho$; hence,

$$\alpha_{r+1} = u_0 + \varepsilon^r \rho + u_2 \varepsilon^{2r} \rho^2 + \cdots + u_{s-1} \varepsilon^{(s-1)r} \rho^{s-1}$$

for $r = 0, 1, \ldots, s-1$ are roots of polynomial (5.1).

For example, for $s = 3$ we get

$$\begin{aligned} \alpha_1 &= u_0 + \rho + u_2 \rho^2, \\ \alpha_2 &= u_0 + \varepsilon \rho + u_2 \varepsilon^2 \rho^2, \\ \alpha_3 &= u_0 + \varepsilon^2 \rho + u_2 \varepsilon \rho^2. \end{aligned}$$

Since $1 + \varepsilon + \varepsilon^2 = 0$, we have

$$\begin{aligned} \alpha_1 + \alpha_2 + \alpha_3 &= 3u_0, \\ \alpha_1 + \varepsilon^{-1} \alpha_2 + \varepsilon^{-2} \alpha_3 &= 3\rho, \\ \alpha_1 + \varepsilon^{-2} \alpha_2 + \varepsilon^{-1} \alpha_3 &= 3u_2 \rho^2. \end{aligned}$$

Therefore, $\rho = \frac{1}{3}\left(\alpha_1 + \varepsilon^2 \alpha_2 + \varepsilon \alpha_3\right)$. For $s > 3$ we get more cumbersome formulas but the arguments remain the same. The proof of the theorem for the last radical ρ_m is completed.

Let us now turn to ρ_{m-1}. We have shown above (for $s = 3$) that the expressions $u_0, \rho, u_2\rho^2, \ldots, u_{s-1}\rho^{s-1}$ can be polynomially expressed in terms of roots $\alpha_1, \ldots, \alpha_n$ of polynomial (5.1). Moreover, they lie in the field $K(\rho_1, \ldots, \rho_{m-1})$, so that each of the values indicated can be represented in the form

$$v_0 + v_1 \rho_{m-1} + v_2 \rho_{m-1}^2 + \cdots + v_{t-1} \rho_{m-1}^{t-1},$$

where $v_i \in K(\rho_1, \ldots, \rho_{m-2})$. The sequence of radicals $\rho_1, \ldots, \rho_m$ is minimal, so that the equations $v_1 = v_2 = \cdots = v_{t-1} = 0$ cannot be simultaneously satisfied for all the quantities because otherwise we could have excluded ρ_{m-1}. Therefore, there exists a relation of the form

$$v_0 + v_1 \rho_{m-1} + v_2 \rho_{m-1}^2 + \cdots + v_{t-1} \rho_{m-1}^{t-1} = r(\alpha_1, \ldots, \alpha_n),$$

where $v_i \in K(\rho_1, \ldots, \rho_{m-2})$, not all elements $v_1, \ldots, v_{t-1}$ vanish and $r(\alpha_1, \ldots, \alpha_n)$ is a polynomial over K. Consider the polynomial

$$G(x) = \prod \left(x - r(\alpha_{\sigma(1)}, \ldots, \alpha_{\sigma(n)})\right),$$

where the product runs over all the permutations $\sigma \in S_n$. The coefficients of G are the symmetric polynomials of the roots of polynomial (5.1), so that they can be polynomially expressed in terms of the coefficients of polynomial (5.1). Thus, G is a polynomial over K and

$$\beta = v_0 + v_1 \rho_{m-1} + \cdots + v_{t-1} \rho_{m-1}^{t-1}$$

is a root of this polynomial. It is also clear that the root β can be expressed by means of the radicals (with the help of the sequence of radicals $\rho_1, \ldots, \rho_{m-1}$). By Theorem 6.5.3 replacing ρ_{m-1} with the radical ρ' of the same degree we may assume that $v_1 = 1$. We can now apply to ρ' the same arguments as we applied to ρ. Iterating the arguments for ρ_{m-2}, and so on, down to ρ_1 completes the proof. □

Now we can pass to the proof of Abel's theorem proper.

6.5.5. THEOREM (Abel). *For $n \geq 5$ it is impossible to express the roots of the general nth degree polynomial in radicals.*

PROOF. Suppose that a certain root α_1 of the general nth degree polynomial

$$x^n + c_1 x^{n-1} + c_2 x^{n-2} + \cdots + c_n$$

can be expressed in radicals. Then by Theorems 6.5.1–6.5.4 there exists an expression of α_1 in radicals of the following particular form. The root α_1 is obtained by consecutively adjoining the radicals $\rho_1, \ldots, \rho_m$ of prime degrees to the ground field, and these radicals, in their turn, are polynomials in the roots $\alpha_1, \ldots, \alpha_n$ of the initial polynomial. More precisely, let $\varepsilon_1, \ldots, \varepsilon_m$ be primitive roots of unity whose degrees are equal to the degrees of the radicals $\rho_1, \ldots, \rho_m$, respectively, $\Delta = \mathbb{Q}(c_1, \ldots, c_n)$, and

$$K = \Delta(\varepsilon_1, \ldots, \varepsilon_m) = \mathbb{Q}(\varepsilon_1, \ldots, \varepsilon_m, c_1, \ldots, c_n).$$

Then α_1 can be polynomially expressed over K in terms of $\rho_1, \ldots, \rho_m$, i.e.,

$$\alpha_1 = r(\rho_1, \ldots, \rho_m, c_1, \ldots, c_n),$$

where r is a polynomial over $\mathbb{Q}(\varepsilon_1, \ldots, \varepsilon_m)$. In their turn, $\rho_1, \ldots, \rho_m$ can be polynomially expressed over K in terms of $\alpha_1, \ldots, \alpha_n$, i.e.,

$$\rho_i = r_i(\alpha_1, \ldots, \alpha_n, c_1, \ldots, c_n),$$

where r_i is a polynomial over $\mathbb{Q}(\varepsilon_1, \ldots, \varepsilon_m)$. Since we deal with the general polynomial of degree n, we may assume that $\alpha_1, \ldots, \alpha_n$ are independent variables and $c_1, \ldots, c_n$ are (up to a sign) the elementary symmetric polynomials of $\alpha_1, \ldots, \alpha_n$.

Let us show that for $n \geq 5$ the assumption on solvability in radicals of the general algebraic equation of degree n leads to a contradiction. To this end consider the permutation

$$T = \begin{pmatrix} 123456\ldots n \\ 234516\ldots n \end{pmatrix}$$

that cyclically permutes the first 5 elements, the others being fixed. Let us prove that under the action of T on the roots $\alpha_1, \ldots, \alpha_n$ the first radical ρ_1 does not change. Since

$$\rho_1 = r_1(\alpha_1, \ldots, \alpha_n, c_1, \ldots, c_n) = \sqrt[p]{a_1},$$

where a_1 is a polynomial of $c_1, \ldots, c_n$ over the field $\mathbb{Q}(\varepsilon_1, \ldots, \varepsilon_m)$, the equation $\rho_1^p = a_1$ can be considered as a relation of the form

$$\varphi(\alpha_1, \ldots, \alpha_n, c_1, \ldots, c_n) = 0,$$

where φ is a polynomial over $\mathbb{Q}(\varepsilon_1, \ldots, \varepsilon_m)$.

Let us show that any relation of this form is preserved under any permutation of the roots $\alpha_1, \ldots, \alpha_n$. Let $\beta_1 = \alpha_{i_1}, \ldots, \beta_n = \alpha_{i_n}$, where $i_1, \ldots, i_n$ is a permutation of the numbers $1, 2, \ldots, n$. Then

$$\varphi(\beta_1, \ldots, \beta_n, d_1, \ldots, d_n) = 0,$$

where $d_i = c_i(\beta_1, \ldots, \beta_n)$. Clearly, $d_i = c_i(\alpha_1, \ldots, \alpha_n) = c_i$ because the functions c_i are symmetric. Hence,

$$\varphi(\alpha_{i_1}, \ldots, \alpha_{i_n}, c_1, \ldots, c_n) = 0.$$

Thus, the relation $\rho_1^p = a_1$ is preserved under the action of T on the roots $\alpha_1, \dots, \alpha_n$, i.e., $T(\rho_1^p) = T(a_1)$. Clearly, $T(\rho_1^p) = T(\rho_1)^p$. Since a_1 only depends on the symmetric functions of roots, $T(a_1) = a_1$. Therefore, $T(\rho_1) = \varepsilon_1^\lambda \rho_1$ and $T^m(\rho_1) = \varepsilon_1^{m\lambda}\rho_1$. But $T^5 = I$ is the identity substitution, hence, $\varepsilon_1^{5\lambda}\rho_1 = T^5(\rho_1) = \rho_1$, i.e., $\varepsilon_1^{5\lambda} = 1$.

Let us now turn to the substitutions

$$U = \begin{pmatrix} 123456\dots n \\ 124536\dots n \end{pmatrix}, \quad V = \begin{pmatrix} 123456\dots n \\ 231456\dots n \end{pmatrix}.$$

It is easy to verify that $U^3 = V^3 = 1$; hence, $U(\rho_1) = \varepsilon_1^\mu \rho_1$ and $V(\rho_1) = \varepsilon_1^\nu \rho_1$, and $\varepsilon_1^{3\mu} = \varepsilon_1^{3\nu} = 1$. Moreover, $UV = T$; hence,

$$T(\rho_1) = VU(\rho_1) = \varepsilon_1^{\mu+\nu}\rho_1.$$

Hence, $\varepsilon_1^\lambda = \varepsilon_1^{\mu+\nu}$ so that $\varepsilon_1^\lambda = \varepsilon_1^{6\lambda}\varepsilon_1^{-5\lambda} = \varepsilon_1^{6(\mu+\nu)} = 1$ because $\varepsilon_1^{5\lambda} = \varepsilon_1^{6\mu} = \varepsilon_1^{6\nu} = 1$. As a result we get $T(\rho_1) = \rho_1$.

Passing consecutively to the radicals $\rho_2, \dots, \rho_m$ we similarly get $T(\rho_i) = \rho_i$ for $i = 2, \dots, m$.

Since $\rho_i = r_i(\alpha_1, \dots, \alpha_n, c_1, \dots, c_n)$, it follows that the equation

$$\alpha_1 = r(\rho_1, \dots, \rho_m, c_1, \dots, c_n)$$

can be considered as a relation between $\alpha_1, \dots, \alpha_n, c_1, \dots, c_n$ over $\mathbb{Q}(\varepsilon_1, \dots, \varepsilon_m)$. This relation is preserved under the action of T, i.e.,

$$T(\alpha_1) = r(T(\rho_1), \dots, T(c_n)) = r(\rho_1, \dots, c_n)$$

since $T(c_i) = c_i$ and $T(\rho_i) = \rho_i$. Therefore, $T(\alpha_1) = \alpha_1$. On the other hand, by the definition of T we get $T(\alpha_1) = \alpha_2$; hence, $\alpha_1 = \alpha_2$. The relation $\alpha_1 = \alpha_2$ contradicts the independence of the roots of the general equation. □

§6.6. The Tschirnhaus transformations. Quintic equations in Bring's form

In 1683 in the journal *Acta Eruditorum* **E.W. von Tschirnhaus**[1] (1651–1708) published a method for transformation of algebraic equations which, Tschirnhaus believed, enabled one to solve in radicals the equation of any degree. **Leibniz** immediately announced that Tschirnhaus' claim on the universality of this transformation was not valid. The catch is that in order to solve a quintic equation with the help of the Tschirnhaus transformations one has to solve an equation of degree 24.

Still, the Tschirnhaus transformation has important applications. For example, with its help any quintic equation without multiple roots can be reduced to the form $y^5 + 5y = a$ and in the process we only have to solve equations of degrees 2 and 3. In Chapter 7 we will show that equations of such a form can then be solved using theta functions.

---

[1]The mathematicians often write Tschirnhausen, but as is clear from the works of historians of mathematics, the correct spelling is *Tschirnhaus*. (Regrettably, his original works were inaccessible for us.) *The authors.*

The Tschirnhaus transformation of the equation $x^n + c_1x^{n-1} + \cdots + c_n = 0$ is as follows. Let $x_1, \ldots, x_n$ be the roots of this equation. Let us consider a rational function φ that is finite at the points $x_1, \ldots, x_n$.

Set $y_i = \varphi(x_i)$ and let

$$y^n + q_1y^{n-1} + \cdots + q_n = 0$$

be the equation for which $y_1, \ldots, y_n$ are the roots. Further on, we will show that if this equation has no multiple roots, then the x_i can be expressed in terms of the y_i. By selecting an appropriate φ we may assure that the coefficients $q_1, \ldots, q_{n-1}$ become zeros. But to do this we have to solve an equation of degree $(n-1)!$ and this was precisely the objection of Leibniz.

Without loss of generality we can use a polynomial of degree not higher than $n-1$ instead of the rational function φ. We can do this because of the following statement.

6.6.1. THEOREM. *Let $x_1, \ldots, x_n$ be the roots of a polynomial f whose degree is equal to n and $\varphi = P/Q$, where P and Q are polynomials such that $Q(x_i) \neq 0$ for $i = 1, \ldots, n$. Then there exists a polynomial g of degree not higher than $n-1$ for which the values of g at points $x_1, \ldots, x_n$ coincide with the values of φ at these points.*

PROOF. By the assumption the polynomials f and Q have no common roots, hence are relatively prime. Therefore there exist polynomials a and b for which $af + bQ = 1$. Since $f(x_i) = 0$, we have $b(x_i) = 1/Q(x_i)$. Hence,

$$\varphi(x_i) = P(x_i)/Q(x_i) = P(x_i)b(x_i).$$

Thus, for g we can take the residue of the division of Pb by f. □

In the sequel we will assume that to the equation

$$f(x) = x^n + c_1x^{n-1} + \cdots + c_n = 0$$

the transformation

$$y = g(x) = p_0 + p_1x + \cdots + p_{n-1}x^{n-1}$$

is applied. Let us show how to calculate in this case the coefficients of the polynomial $y^n + q_1y^{n-1} + \cdots + q_n$ whose roots are $y_i = g(x_i), i = 1, \ldots, n$. For simplicity, we confine ourselves to the case $n = 3$.

If $x^3 = -c_1x^2 - c_2x - c_3$, then

$$yx = p_0x + p_1x^2 + p_2\left(-c_1x^2 - c_2x - c_3\right) = p_0' + p_1'x + p_2'x^2.$$

Similarly, $yx^2 = p_0'' + p_1''x + p_2''x^2$, where the p_i'' are linear functions of the parameters p_i. Therefore, if x_i is a root of f and $y_i = g(x_i)$, then the system of equations

$$\text{(6.1)} \qquad \begin{cases} (p_0 - y)z_0 + p_1z_1 + p_2z_2 = 0, \\ p_0'z_0 + (p_1' - y)z_1 + p_2'z_2 = 0, \\ p_0''z_0 + p_1''z_1 + (p_2'' - y)z_2 = 0 \end{cases}$$

has a nonzero solution $(z_0, z_1, z_2) = (1, x_i, x_i^2)$. Set

$$A = \begin{pmatrix} p_0 & p_1 & p_2 \\ p_0' & p_1' & p_2' \\ p_0'' & p_1'' & p_2'' \end{pmatrix}.$$

Then $\det(A - yI) = 0$ for $y_i = g(x_i)$. If the polynomial $\det(yI - A)$ has no multiple roots, it coincides with the desired polynomial $y^n + q_1 y^{n-1} + \cdots + q_n$. Since the elements of A linearly depend on the parameters p_i, the coefficient q_k is a kth degree polynomial of the parameters p_i.

If $y^n + q_1 y^{n-1} + \cdots + q_n$ has no multiple roots, A has no multiple eigenvalues. Therefore, to each eigenvalue of A there corresponds a unique (up to proportionality) solution of system (6.1). This means that from a root y_i of the transformed polynomial a root x_i of the initial polynomial is uniquely recovered and that the root x_i can be rationally expressed in terms of the root y_i.

The Tschirnhaus transformation enables one to solve in radicals the equations of degrees 3 and 4. A cubic equation can be reduced to the form $y^3 + q_3 = 0$ after we solve the system depending on the parameters p_0, p_1, p_2 consisting of a linear equation $q_1 = 0$ and a second degree equation $q_2 = 0$. To do this, we have to solve a quadratic equation.

A fourth degree equation can be reduced to the form

$$y^4 + q_2 y^2 + q_4 = 0.$$

To do so, we have to solve a system consisting of a linear equation $q_1 = 0$ and a third degree equation $q_3 = 0$, which reduces to solving a cubic equation.

A fifth degree equation can be reduced to the form

$$y^5 + q_4 y + q_5 = 0$$

after the system of equations $q_1 = q_2 = q_3 = 0$ is solved. To this end we have to solve an equation of degree 6. A more detailed analysis was performed in 1789 by the Swedish lawyer, historian and mathematician **Bring**, who demonstrated that in this case instead of a sixth degree equation it suffices to solve equations of degrees 2 and 3. Indeed, in order to satisfy the equation $q_1 = 0$, express one of the parameters $p_0, \ldots, p_4$ as a linear function of the other parameters. Then the coefficient q_2 is a quadratic form with respect to four of the parameters p_i. This quadratic form can be reduced to the form $u_1^2 + u_2^2 - v_1^2 - v_2^2$, where u_j and v_j are linear functions of the p_i (to perform the reduction, we have to calculate square roots). To satisfy the equation $q_2 = 0$, it suffices to solve the system of linear equations $u_1 = v_1, u_2 = v_2$.

This leaves us two parameters and the equation $q_3 = 0$ for them is a third degree equation. As a result, we get an equation of the form $y^5 + q_4 y + q_5 = 0$.

If $q_4 \neq 0$, then with the help of a linear change of variables we can reduce this equation to the form $y^5 + 5y = a$.

CHAPTER 7

Theta Functions and Solutions of Quintic Equations

§7.1. Definition of theta functions

Theta functions are entire functions with one genuine period and one *quasiperiod* (though under addition of the quasiperiod to the argument the function changes, but this change is subject to a sufficiently simple law).

Setting $q = e^{i\pi\tau}$ one can map the upper half plane $H = \{\tau \in \mathbb{C} \mid \operatorname{Im}\tau > 0\}$ in the interior of the unit disk $D = \{q \in \mathbb{C} \mid |q| \leq 1\}$. Indeed, let $\tau = x + iy$, where $x, y \in \mathbb{R}$. Then $q = e^{i\pi x - \pi y} = e^{-\pi y}e^{i\pi x}$ and $|q| = e^{-\pi y}$. Hence, $|q| < 1$ if and only if $y > 0$, i.e., $\operatorname{Im}\tau > 0$.

Fix a number $\tau \in H$ and consider the series

$$\Theta_3(v \mid \tau) = \sum_{m=-\infty}^{\infty} e^{(m^2\tau+2mv)\pi i} = \sum_{m=-\infty}^{\infty} q^{m^2}e^{2\pi imv}.$$

The absolute value of the ratio of the consecutive terms of this series is equal to $\left|q^{2m+1}e^{2\pi iv}\right| \leq |q|^{2m+1}e^{2\pi|v|}$. Since $\lim_{m\to\infty}|q|^{2m+1} = 0$, it follows that $\Theta_3(v \mid \tau)$ is a series of entire functions of v converging uniformly in the domain $|v| \leq c$, where c is a constant. Therefore, $\Theta_3(v \mid \tau)$ is itself an entire function of v. For brevity we will often denote this function by $\Theta_3(v)$.

By replacing v with $v+1$ we get the same series, since $e^{2\pi im(v+1)} = e^{2\pi imv}$. This means that $\Theta_3(v+1) = \Theta_3(v)$. Moreover, $\Theta_3(v+\tau) = A\Theta_3(v)$, where $A = q^{-1}e^{-2\pi iv}$. Indeed,

$$\begin{aligned}\Theta_3(v+\tau) &= \sum q^{m^2}e^{2\pi imv}q^{2m}\\ &= q^{-1}e^{-2\pi iv}\sum q^{(m+1)^2}e^{2\pi i(m+1)} = q^{-1}e^{-2\pi iv}\Theta_3(v).\end{aligned}$$

It is also convenient to consider the functions

$$\begin{aligned}\Theta_0(v) &= \Theta_3\left(v+\frac{1}{2}\right) = \sum q^{m^2}e^{2\pi imv}e^{\pi im} = \sum(-1)^m q^{m^2}e^{2\pi imv};\\ \Theta_1(v) &= ie^{-\pi i\left(v-\frac{\tau}{4}\right)}\Theta_3\left(v+\frac{1-\tau}{2}\right)\\ &= ie^{-\pi i\left(v-\frac{\tau}{4}\right)}\sum q^{m^2}e^{2\pi imv}e^{\pi im}e^{-\pi im\tau}\\ &= i\sum(-1)^m q^{\left(m-\frac{1}{2}\right)^2}e^{\pi i(2m-1)v}\end{aligned}$$

(the function q^λ is multivalued for $\lambda \notin \mathbb{Z}$; we take $q^\lambda = e^{i\pi\lambda\tau}$); and

$$\begin{aligned}\Theta_2(v) &= e^{-\pi i\left(v-\frac{\tau}{4}\right)}\Theta_3\left(v-\frac{\tau}{2}\right) = e^{-\pi i\left(v-\frac{\tau}{4}\right)}\sum q^{m^2}e^{2\pi imv}e^{-\pi imv}\\ &= \sum q^{\left(m-\frac{1}{2}\right)^2}e^{\pi i(2m-1)v}.\end{aligned}$$

The functions Θ_0, Θ_1, Θ_2 and Θ_3 are called *theta functions.*

Problems

7.1.1. Prove that $\Theta_k(v+1) = \Theta_k(v)$ for $k = 0, 3$ and $\Theta_k(v+1) = -\Theta_k(v)$ for $k = 1, 2$.

7.1.2. Prove that $\Theta_k(v+\tau) = A\Theta_k(v)$ for $k = 2, 3$ and $\Theta_k(v+\tau) = -A\Theta_k(v)$ for $k = 0, 1$.

§7.2. Zeros of theta functions

Let the numbers m and k be related by the formula $2m-1 = 1-2k$, i.e., $k = 1-m$. Then $(-1)^m = -(-1)^k$ and $q^{\left(m-\frac{1}{2}\right)^2} = q^{\left(k-\frac{1}{2}\right)^2}$, so that $\Theta_1(-v) = -\Theta_1(v)$. In particular, $\Theta_1(0) = 0$.

Since $\Theta_1(v+1) = -\Theta_1(v)$ and $\Theta_1(v+\tau) = -A\Theta_1(v)$, we have $\Theta_1(m+n\tau) = 0$. Let us show that the function Θ_1 has no other zeros. It suffices to verify that the function Θ_1 has only one zero inside the parallelogram Π with vertices $\frac{\pm 1 \pm \tau}{2}$.

Clearly,

$$\begin{aligned}\frac{\Theta_1'(v+1)}{\Theta_1(v+1)} &= \frac{\Theta_1'(v)}{\Theta_1(v)};\\ \frac{\Theta_1'(v+\tau)}{\Theta_1(v+\tau)} &= \frac{-A'(v)\Theta_1(v) - A(v)\Theta_1'(v)}{-A(v)\Theta_1(v+1)}\\ &= \frac{A'(v)}{A(v)} + \frac{\Theta_1'(v)}{\Theta_1(v)} = -2\pi i + \frac{\Theta_1'(v)}{\Theta_1(v)}.\end{aligned}$$

Therefore, if C is the boundary of Π oriented counterclockwise, then

$$\frac{1}{2\pi i}\int_C \frac{\Theta_1'(z)}{\Theta_1(z)}dz = 1.$$

Indeed, the sum of the integrals along the horizontal sides is equal to $2\pi i$, since the lengths of these sides are equal to 1, whereas the integrals along the other two sides of Π cancel.

If $f(z) = c_k(z-a_k)^k + \cdots$, then $\frac{f'(z)}{f(z)} = \frac{k}{z-a_k} + \cdots$. Therefore, for an entire function f the value

$$\frac{1}{2\pi i}\int_C \frac{f'(z)}{f(z)}dz$$

is equal to the number of zeros (multiplicities counted) inside C. Therefore, inside Π, the function Θ_1 has exactly one zero.

Starting from the expressions for the functions Θ_0, Θ_1 and Θ_2 in terms of Θ_3, we can easily get the following table for zeros of theta functions:

function	$\Theta_0(v)$	$\Theta_1(v)$	$\Theta_2(v)$	$\Theta_3(v)$
its zeros	$m+\left(n+\frac{1}{2}\right)\tau$	$m+n\tau$	$m+\frac{1}{2}+n\tau$	$m+\frac{1}{2}+\left(n+\frac{1}{2}\right)\tau$

Problems

7.2.1. Prove that the functions Θ_0, Θ_2 and Θ_3 are even.

§7.3. The relation $\Theta_3^4 = \Theta_2^4 + \Theta_0^4$

The quantities $\Theta_i = \Theta_i(0)$ for $i = 0, 2, 3$ and $\Theta_1' = \Theta_1'(0)$ are called *theta constants*. Recall that they depend on the parameter τ.

Consider the function

$$f(v) = \frac{a\Theta_2^2(v) + b\Theta_3^2(v)}{\Theta_0^2(v)}, \quad \text{where } a, b \in \mathbb{C}.$$

The numbers 1 and τ are periods of f; hence, f is a doubly periodic function with fundamental parallelogram Π. The fundamental parallelogram can be shifted so that only one zero of the function $\Theta_0(v)$ lies inside it; assume that it is $\frac{\tau}{2}$. If we select numbers a and b so that $a\Theta^2(\frac{\tau}{2}) + b\Theta_3^2(\frac{\tau}{2}) = 0$, then, inside Π, the elliptic function $f(v)$ has a pole of multiplicity not greater than 1 and, therefore, $f(v)$ is a constant.

Substituting $v = \frac{\tau}{2}$ and $v = 0$, respectively, in the relation

$$\Theta_2(v) = e^{-\pi i\left(v-\frac{\tau}{4}\right)}\Theta_3\left(v - \frac{\tau}{2}\right)$$

we get

$$\Theta_2\left(\frac{\tau}{2}\right) = e^{-\frac{\pi i\tau}{4}}\Theta_3(0), \quad \text{resp.} \quad \Theta_2(0) = e^{\frac{\pi i\tau}{4}}\Theta_3\left(-\frac{\tau}{2}\right) = e^{\frac{\pi i\tau}{4}}\Theta_3\left(\frac{\tau}{2}\right).$$

Hence, $a\Theta_2^2(\frac{\tau}{2}) + b\Theta_3^2(\frac{\tau}{2}) = aB^2\Theta_3^2 + bB^2\Theta_2^2$, where $B = e^{-\frac{\pi i\tau}{4}}$. Set $a = -\Theta_2^2$ and $b = \Theta_3^2$. Then $aB^2\Theta_3^2 + bB^2\Theta_2^2 = 0$ and

$$-\Theta_2^2(v)\Theta_2^2 + \Theta_3^2(v)\Theta_3^2 = c\Theta_0^2(v).$$

To compute c, set $v = \frac{1}{2}$. The function Θ_2 vanishes at this point, so to calculate $\Theta_3(\frac{1}{2})$ and $\Theta_0(\frac{1}{2})$ we can substitute $v = 0$ and $v = \frac{1}{2}$ in the relation $\Theta_0(v) = \Theta_3(v + \frac{1}{2})$. We get $\Theta_0 = \Theta_3(\frac{1}{2})$ and $\Theta_0(\frac{1}{2}) = \Theta_3(1) = \Theta_3$. Therefore,

$$\Theta_0^2\Theta_3^2 = c\Theta_3^2,$$

i.e., $c = \Theta_0^2$. Thus,

$$\Theta_3^2(v)\Theta_3^2 - \Theta_2^2(v)\Theta_2^2 = \Theta_0^2(v)\Theta_0.$$

In particular, for $v = 0$ we get

$$\Theta_3^4 = \Theta_2^4 + \Theta_0^4,$$

i.e.,

$$(1+2q+2q^4+2q^9+\cdots)^4 = 16q(1+q^{1\cdot 2}+q^{2\cdot 3}+q^{3\cdot 4}+\cdots)^4+(1-2q+2q^4-2q^9+\cdots)^4.$$

§7.4. Representation of theta functions by infinite products

The numbers $m + \frac{1}{2} + \left(n + \frac{1}{2}\right)\tau$ are zeros of the function

$$\Theta_3(v) = \sum_{k=-\infty}^{\infty} q^{k^2} e^{2\pi i k v}.$$

Set $s = e^{2\pi i v}$. This substitution sends the zeros of $\Theta_3(v)$ to the points

$$e^{2\pi i\left(m+\frac{1}{2}\right)} e^{2\pi i\left(n+\frac{1}{2}\tau\right)} = -q^{2n+1}, \qquad \text{where } q = e^{\pi i \tau}.$$

Let us put these points into the two sets:

$$(4.1) \qquad -q^{-1}, -q^{-3}, -q^{-5}, \dots,$$

$$(4.2) \qquad -q^{1}, -q^{3}, -q^{5}, \dots.$$

The limit point of the set (4.1) is ∞ and the limit point of the set (4.2) is 0.

The series $\sum_{k=1}^{\infty} \left|q^{2k-1}\right|$ converges; hence, the function

$$f_1(s) = \prod_{k=1}^{\infty} \left(1 + q^{2k-1}s\right)$$

is an entire function of s with zeros at the points (4.1). Similarly, the function

$$f_2(s) = \prod_{k=1}^{\infty} \left(1 + q^{2k-1}s^{-1}\right)$$

is an entire function if we consider it as a function of s^{-1} (as a function of s it has a singularity at the origin). The zeros of $f_2(s)$ are the points (4.2).

Consider the function $f(s) = f_1(s)f_2(s)$. The function $g(v) = f(e^{2\pi i v}) = f(s)$ has the same zeros as the function $\Theta_3(v)$. Under the change of variables $v \mapsto v + 1$ the value $s = e^{2\pi i v}$ does not change and, therefore, $g(v + 1) = g(v)$. The change of variables $v \mapsto v + \tau$ replaces s with $e^{2\pi i v}e^{2\pi i \tau} = sq^2$; hence,

$$g(v + \tau) = \frac{1 + q^{-1}s^{-1}}{1 + qs} g(v) = q^{-1}e^{-2\pi i v} g(v).$$

Therefore, the ratio of the functions $\Theta_3(v)$ and $g(v)$ is an entire doubly periodic function, i.e., a constant. Therefore,

$$(4.3) \qquad \Theta_3(v) = c \prod_{k=1}^{\infty} \left(1 + q^{2k-1}e^{2\pi i v}\right)\left(1 + q^{2k-1}e^{-2\pi i v}\right).$$

Similar expansions can be obtained for the remaining theta functions. Since $e^{2\pi i\left(v+\frac{1}{2}\right)} = e^{-2\pi i v}$, it follows that

$$(4.4) \qquad \Theta_0(v) = \Theta_3\left(v + \frac{1}{2}\right) = c \prod_{k=1}^{\infty} \left(1 - q^{2k-1}e^{2\pi i v}\right)\left(1 - q^{2k-1}e^{-2\pi i v}\right).$$

Since $q^{2k-1}e^{2\pi i\left(v+\frac{1}{2}-\frac{\tau}{2}\right)} = -q^{2k-2}e^{2\pi iv}$ and $q^{2k-1}e^{-2\pi i\left(v+\frac{1}{2}-\frac{\tau}{2}\right)} = -q^{2k}e^{-2\pi iv}$, we have

$$\begin{aligned}\Theta_1(v) &= ie^{-\pi iv}q^{\frac{1}{4}}\Theta_3\left(v+\frac{1}{2}-\frac{\tau}{2}\right)\\ &= i\left(1-e^{2\pi iv}\right)e^{-\pi iv}q^{\frac{1}{4}}c\prod_{k=1}^{\infty}\left(1-q^{2k}e^{2\pi iv}\right)\left(1-q^{2k}e^{-2\pi iv}\right).\end{aligned}$$

It is also clear that $i\left(1-e^{2\pi iv}\right)e^{-\pi iv} = 2\sin\pi v$, so that

$$\Theta_1(v) = 2(\sin\pi v)\,q^{\frac{1}{4}}c\prod_{k=1}^{\infty}\left(1-q^{2k}e^{2\pi iv}\right)\left(1-q^{2k}e^{-2\pi iv}\right). \tag{4.5}$$

It is easy to verify that $\Theta_2(v) = \Theta_1(v+\frac{1}{2})$; hence,

$$\Theta_2(v) = 2(\cos\pi v)\,q^{\frac{1}{4}}c\prod_{k=1}^{\infty}\left(1+q^{2k}e^{2\pi iv}\right)\left(1+q^{2k}e^{-2\pi iv}\right). \tag{4.6}$$

Now, let us prove that

$$c = \prod_{k=1}^{\infty}\left(1-q^{2k}\right). \tag{4.7}$$

Consider the sequence of functions

$$F_n(s) = \prod_{k=1}^{n}\left(1-q^{2k-1}s\right)\left(1-q^{2k-1}s^{-1}\right) = \sum_{k=-n}^{n}a_k(n)s^k.$$

This sequence converges uniformly to the function

$$\prod_{k=1}^{\infty}\left(1-q^{2k-1}s\right)\left(1-q^{2k-1}s^{-1}\right) = \frac{1}{c}\Theta_0(v) = \frac{1}{c}\sum_{k=-\infty}^{\infty}(-1)^kq^{k^2}s^k.$$

Comparing the coefficients of degree zero in s, we get $\frac{1}{c} = \lim_{n\to\infty}a_0(n)$.

Clearly, $a_n(n) = (-1)^nq^{1+3+\cdots+(2n-1)} = (-1)^nq^{n^2}$. Moreover,

$$\frac{F_n(q^2s)}{F_n(s)} = \frac{\left(1-q^{2n+1}s\right)\left(1-q^{-1}s^{-1}\right)}{(1-qs)\left(1-q^{2n-1}s^{-1}\right)} = -\frac{1-q^{2n+1}s}{qs-q^{2n}};$$

hence,

$$\left(qs-q^{2n}\right)\sum_{k=-n}^{n}a_k(n)q^{2k}s^k = -\left(1-q^{2n+1}s\right)\sum_{k=-n}^{n}a_k(n)s^k,$$

i.e.,

$$\sum_{k=-n}^{n}a_k(n)\left(1-q^{2(n+k)}\right)s^k = \sum_{k=-n}^{n}a_k(n)\left(q^{2n+1}-q^{2k+1}\right)s^{k+1}.$$

Therefore,

$$a_0(n) = q^{n^2}\frac{\prod_{k=1}^{n}\left(1-q^{2(n+k)}\right)}{\prod_{k=1}^{n}\left(q^{2k+1}-q^{2n+1}\right)} = \frac{\prod_{k=1}^{n}\left(1-q^{2(n+k)}\right)}{\prod_{k=1}^{n}\left(1-q^{2k}\right)}.$$

Let $|q|^2 = \alpha < 1$. Then $1 - \alpha^n \leq \left|1 - q^{2(n+k)}\right| \leq 1 + \alpha^n$. Moreover, $\lim_{n\to\infty} n \ln(1 \pm \alpha^n) = 0$. Hence $\lim_{n\to\infty} \prod_{k=1}^{n} \left(1 - q^{2(n+k)}\right) = 1$ and

$$c = \lim_{n\to\infty} 1/a_0(n) = \prod_{k=1}^{\infty} \left(1 - q^{2k}\right).$$

§7.5. The relation $\Theta_1'(0) = \pi\Theta_0(0)\Theta_2(0)\Theta_3(0)$

Formulas (4.4)–(4.7) imply that

(5.1)
$$\Theta_0(0) = c \prod_{k=1}^{\infty} \left(1 - q^{2k-1}\right)^2;$$

(5.2)
$$\Theta_2(0) = 2q^{\frac{1}{4}} c \prod_{k=1}^{\infty} \left(1 + q^{2k}\right)^2;$$

(5.3)
$$\Theta_3(0) = c \prod_{k=1}^{\infty} \left(1 + q^{2k-1}\right)^2;$$

(5.4)
$$\Theta_1'(0) = 2\pi q^{\frac{1}{4}} c \prod_{k=1}^{\infty} \left(1 - q^{2k}\right)^2.$$

To prove the last formula, it suffices to observe that $\Theta_1'(0) = \lim_{v\mapsto 0} \frac{\Theta_1(v)}{v}$.

Since $\prod_{k=1}^{\infty} \left(1 - q^{2k}\right)^2 = c^2$, we have $\Theta_1'(0) = 2\pi q^{\frac{1}{4}} c^3$. Therefore, to prove the relation $\Theta_1'(0) = \pi\Theta_0(0)\Theta_2(0)\Theta_3(0)$, it suffices to verify that

$$\prod_{k=1}^{\infty} \left(1 - q^{2k-1}\right)\left(1 + q^{2k}\right)\left(1 + q^{2k-1}\right) = 1.$$

Clearly,

$$\prod_{k=1}^{\infty} \left(1 - q^{2k-1}\right) = \prod_{n=1}^{\infty} \left(1 - q^n\right)\left(1 - q^{2n}\right)^{-1};$$
$$\prod_{k=1}^{\infty} \left(1 + q^{2k}\right)\left(1 + q^{2k-1}\right) = \prod_{n=1}^{\infty} \left(1 + q^n\right).$$

Therefore,

$$\begin{aligned}\prod_{k=1}^{\infty} \left(1 - q^{2k-1}\right)\left(1 + q^{2k}\right)\left(1 + q^{2k-1}\right) &= \prod_{n=1}^{\infty} \left(1 - q^n\right)\left(1 + q^n\right)\left(1 - q^{2n}\right)^{-1} \\ &= \prod_{n=1}^{\infty} \left(1 - q^{2n}\right)\left(1 - q^{2n}\right)^{-1} = 1.\end{aligned}$$

§7.6. Dedekind's η-function and the functions f, f_1, f_2

Formulas (5.1)–(5.4) together with (4.7) enable us to represent the theta constants in the following form:[1]

$$\Theta_1'(0) = 2\pi\eta^3(\tau), \qquad \eta(\tau) = q^{\frac{1}{12}} \prod_{k=1}^{\infty} \left(1 - q^{2k}\right);$$

$$\Theta_3(0) = \eta(\tau) f^2(\tau), \quad f(\tau) = q^{-\frac{1}{24}} \prod_{k=1}^{\infty} \left(1 + q^{2k-1}\right);$$

$$\Theta_0(0) = \eta(\tau) f_1^2(\tau), \quad f_1(\tau) = q^{-\frac{1}{24}} \prod_{k=1}^{\infty} \left(1 - q^{2k-1}\right);$$

$$\Theta_2(0) = \eta(\tau) f_2^2(\tau), \quad f_2(\tau) = \sqrt{2} q^{\frac{1}{12}} \prod_{k=1}^{\infty} \left(1 + q^{2k}\right).$$

The relation $\Theta_3^4(0) = \Theta_2^4(0) + \Theta_0^4(0)$ implies that

(6.1) $$f^8 = f_1^8 + f_2^8.$$

In §7.5 we have shown that

$$\prod_{k=1}^{\infty} \left(1 - q^{2k-1}\right)\left(1 + q^{2k}\right)\left(1 + q^{2k-1}\right) = 1.$$

Hence,

(6.2) $$f f_1 f_2 = \sqrt{2}.$$

The functions f, f_1 and f_2 can be expressed in terms of η as follows:

(6.3) $$f(\tau) = \frac{e^{-\frac{\pi i}{24}} \eta\left(\frac{\tau+1}{2}\right)}{\eta(\tau)};$$

(6.4) $$f_1(\tau) = \frac{\eta\left(\frac{\tau}{2}\right)}{\eta(\tau)};$$

(6.5) $$f_2(\tau) = \sqrt{2}\frac{\eta(2\tau)}{\eta(\tau)}.$$

Let us prove, for instance, formula (6.3). It is easy to verify that

$$\begin{aligned}
\frac{\eta\left(\frac{\tau+1}{2}\right)}{\eta(\tau)} &= q^{\frac{1}{24} - \frac{1}{12}} \left(e^{\frac{\pi i}{2}}\right)^{\frac{1}{12}} \prod_{k=1}^{\infty} \frac{1 - \left(i q^{\frac{1}{2}}\right)^{2k}}{1 - q^{2k}} \\
&= q^{-\frac{1}{24}} e^{\frac{\pi i}{24}} \frac{(1+q)\left(1-q^2\right)\left(1+q^3\right)\left(1-q^4\right)\dots}{\left(1-q^2\right)\left(1-q^4\right)\dots} \\
&= e^{\frac{\pi i}{24}} q^{-\frac{1}{24}} \prod_{k=1}^{\infty} \left(1 - q^{2k-1}\right).
\end{aligned}$$

It is even easier to prove formulas (6.4) and (6.5).

---

[1] Observe that the functions $f(\tau)$, $f_1(\tau)$, $f_2(\tau)$ introduced below and the functions $f(s)$, $f_1(s)$, $f_2(s)$ from §7.4 are different functions.

§7.7. Transformations of theta functions induced by transformations of τ

In §7.1 we have seen how the functions $\Theta_i(v|\tau)$ behave under the replacement of v with $v+1$ or $v+\tau$. It turns out that under the replacement of τ with $\tau+1$ or $-\frac{1}{\tau}$ the functions $\Theta_i(v|\tau)$ are also transformed according to relatively simple laws. For the change of parameter $\tau \mapsto \tau+1$ this is no wonder, since under such a change of parameter $q = e^{\pi i\tau}$ turns into $q' = -q$. Therefore,

$$\Theta_0(v|\tau+1) = \Theta_3(v|\tau), \quad \Theta_3(v|\tau+1) = \Theta_0(v|\tau),$$

$$\Theta_i(v|\tau+1) = e^{\frac{\pi i}{4}}\Theta_i(v|\tau) \quad \text{for } i = 1, 2.$$

The change of parameter $\tau \mapsto \tau' = -\frac{1}{\tau}$, however, sends $q = e^{\pi i\tau}$ into $q' = e^{-\pi i/\tau} = \exp(-\pi^2/\ln q)$. It is surprising that there is a transformation law for the theta function for such a change of parameter. To find this law consider the function

$$g(v) = e^{\pi i\tau' v^2}\frac{\Theta_3(\tau' v|\tau')}{\Theta_3(v|\tau)}.$$

Simple calculations show that $g(v+1) = g(v)$ and $g(v+\tau) = g(v)$, i.e., g is a doubly periodic function. The zeros of the denominator are of the form $v = \left(m+\frac{1}{2}\right)\tau + \left(n+\frac{1}{2}\right)$ and the zeros of the numerator are determined from the relation $\tau' v = \left(m+\frac{1}{2}\right)\tau' + \left(n+\frac{1}{2}\right)$, i.e.,

$$v = \left(m+\frac{1}{2}\right) - \left(n+\frac{1}{2}\right)\tau.$$

Therefore, the zeros of the numerator coincide with the zeros of the denominator; hence, g is an entire doubly periodic function. Therefore, $g = A$, where A is a constant.

By replacing v consecutively with $v+\frac{1}{2}$, $v+\frac{\tau}{2}$ and $v+\frac{1+\tau}{2}$, we derive from the equation

$$\Theta_3(\tau' v|\tau') = Ae^{-\pi i\tau' v^2}\Theta_3(v|\tau) \tag{7.1}$$

the following equations:

$$\Theta_2(\tau' v|\tau') = Ae^{-\pi i\tau' v^2}\Theta_0(v|\tau), \tag{7.2}$$

$$\Theta_0(\tau' v|\tau') = Ae^{-\pi i\tau' v^2}\Theta_2(v|\tau), \tag{7.3}$$

$$\Theta_1(\tau' v|\tau') = iAe^{-\pi i\tau' v^2}\Theta_1(v|\tau). \tag{7.4}$$

Taking the product of equations (7.1)–(7.3) and setting $v = 0$ we get

$$\Theta_2(0|\tau')\Theta_3(0|\tau')\Theta_0(0|\tau') = A^3\Theta_2(0|\tau)\Theta_3(0|\tau)\Theta_0(0|\tau). \tag{7.5}$$

The derivative of (7.4) with respect to v yields at $v = 0$

$$\tau'\Theta_1'(0|\tau') = iA\Theta_1'. \tag{7.6}$$

Since $\Theta_1' = \pi\Theta_0\Theta_2\Theta_3$, equations (7.5) and (7.6) imply that $A^2 = -i\tau$, i.e., $A = \pm\sqrt{-i\tau}$. As a result we get

$$\Theta_3(0|\tau') = \pm\sqrt{-i\tau}\Theta_3(0|\tau).$$

It is also clear that for a purely imaginary τ both $\Theta_3(0|\tau)$ and $\Theta_3(0|\tau')$ are positive. Therefore,

$$\Theta_3\left(0|-\frac{1}{\tau}\right) = \sqrt{-i\tau}\Theta_3(0|\tau). \tag{7.7}$$

Transformations of Dedekind's η-function. Transformations of the theta functions as the parameter τ varies lead to transformations of Dedekind's η-function since

$$2\pi\eta^3(\tau) = \Theta_1'(0|\tau).$$

From the relation $\Theta_1(0|\tau+1) = e^{\frac{\pi i}{4}}\Theta_1(0|\tau)$ we get

$$\eta(\tau+1) = e^{\frac{\pi i}{12}}\eta(\tau). \tag{7.8}$$

If we take into account that $A = \sqrt{-i\tau}$ and $\tau' = -\frac{1}{\tau}$, then (7.6) takes the form

$$\Theta_1'\left(0|-\frac{1}{\tau}\right) = (\sqrt{-i\tau})^3\Theta_1'(0|\tau).$$

Hence,

$$\eta\left(-\frac{1}{\tau}\right) = \sqrt{-i\tau}\,\eta(\tau). \tag{7.9}$$

With the help of formulas (7.8) and (7.9) we can obtain the transformation laws for the functions f, f_1 and f_2 under the change of parameter $\tau \mapsto \tau+1$ or $\tau \mapsto -\frac{1}{\tau}$. For example,

$$f(\tau+1) = \frac{e^{-\frac{\pi i}{24}}\eta\left(\frac{\tau+2}{2}\right)}{\eta(\tau+1)} = \frac{e^{-\frac{\pi i}{24}}\eta\left(\frac{\tau}{2}\right)}{\eta(\tau)} = e^{-\frac{\pi i}{24}}f_1(\tau). \tag{7.10}$$

Similarly, we get

$$f_1(\tau+1) = e^{-\frac{\pi i}{24}}f(\tau), \tag{7.11}$$

$$f_2(\tau+1) = e^{\frac{\pi i}{12}}f_2(\tau). \tag{7.12}$$

Observe that $f(\tau+2) = e^{-\frac{\pi i}{12}}f(\tau)$ and, therefore, $f(\tau+48) = f(\tau)$.

It is easy to verify that

$$f_1\left(-\frac{1}{\tau}\right) = \frac{\eta(-\frac{1}{2\tau})}{\eta(-\frac{1}{\tau})} = \frac{\sqrt{-2i\tau}\eta(2\tau)}{\sqrt{-i\tau}\eta(\tau)} = f_2(\tau). \tag{7.13}$$

Replacing τ with $-\frac{1}{\tau}$ we get

$$f_2\left(-\frac{1}{\tau}\right) = f_1(\tau). \tag{7.14}$$

Similar calculations do not yield an expression for $f(-\frac{1}{\tau})$. However, if we use the fact that

$$f(\tau)f_1(\tau)f_2(\tau) = \sqrt{2}$$

and

$$f\left(-\frac{1}{\tau}\right) f_1(\tau) f_2(\tau) = f\left(-\frac{1}{\tau}\right) f_2\left(-\frac{1}{\tau}\right) f_1\left(-\frac{1}{\tau}\right) = \sqrt{2}$$

we get

$$f\left(-\frac{1}{\tau}\right) = f(\tau). \tag{7.15}$$

Now we prove the relation

$$f(\tau) f\left(\frac{\tau - 1}{\tau + 1}\right) = \sqrt{2}. \tag{7.16}$$

Equations (6.4) and (6.5) imply that $f_1(2\tau)f_2(\tau) = \sqrt{2}$. Since

$$f_1(2\tau) = e^{-\frac{\pi i}{24}} f(2\tau - 1)$$

and

$$f_2(\tau) = f_1\left(-\frac{1}{\tau}\right) = e^{\frac{\pi i}{24}} f\left(-\frac{1}{\tau}\right),$$

then $f(2\tau - 1)f(-\frac{1}{\tau}) = \sqrt{2}$. Set $x = 2\tau - 1$. Then $1 - \frac{1}{\tau} = \frac{x-1}{x+1}$ and the obtained relation is equivalent to (7.16).

§7.8. The general scheme of solution of quintic equations

The remaining part of this chapter is directly concerned with the solutions of quintic equations. The construction is rather complicated and is based on various facts whose proofs often require cumbersome calculations. To help the reader grasp this construction we first describe the general scheme: what is done and to what purpose.

Let $f(\tau)$, $f_1(\tau)$, $f_2(\tau)$ be the functions defined in §7.6. Set

$$u = f(\tau), \quad v_c = f\left(\frac{\tau + c}{5}\right) \quad \text{and} \quad v_\infty = f(5\tau).$$

The study of the behavior of u and v under the changes of parameter $\tau \mapsto \tau + 2$, $\tau \mapsto -\tau^{-1}$ and $\tau \mapsto \frac{\tau-1}{\tau+1}$ shows that these changes transform uv and u/v as follows:

	uv	$\frac{u}{v}$
$\tau \mapsto \tau + 2$	$e^{-\pi i/2}uv$	$e^{\pi i/3}\frac{u}{v}$
$\tau \mapsto -\frac{1}{\tau}$	uv	$\frac{u}{v}$
$\tau \mapsto \frac{\tau-1}{\tau+1}$	$-\frac{2}{uv}$	$-\frac{v}{u}$

The transformation law of the index c of the function v_c is the same as that of τ. Now it is not difficult to demonstrate that

$$\left(\frac{u}{v}\right)^3 + \left(\frac{v}{u}\right)^3 = (uv)^2 - \frac{4}{(uv)^2}$$

or, equivalently,

$$u^6 - u^5v^6 + 4uv + u^6 = 0.$$

Let us consider $u = f(\tau)$ as a parameter. For every value of this parameter we get a sixth degree equation whose roots are explicitly expressed in terms of τ. We would like to do the same for the equation $y^5+5y=a$. It turns out that this can be done as follows. Consider the fifth degree polynomial with the roots

$$w_z=\frac{(v_\infty-v_z)(v_{z+1}-v_{z-1})(v_{z+2}-v_{z-1})}{\sqrt[5]{u^3}},\quad \text{where } z=0,1,2,3,4. \tag{8.1}$$

Calculations will show that this polynomial is of the form

$$w(w^2+5)^2-u^{12}+64u^{-12}.$$

After the substitution

$$y(\tau)=\frac{f_1^8(\tau)-f_2^8(\tau)}{f^2(\tau)(w^2(\tau)+5)} \tag{8.2}$$

we get the equation

$$y^5+5y=\frac{f_1^8-f_2^8}{f^2}.$$

Thus, to solve the equation $y^5+5y=a$, we have to proceed as follows. First, let us find a τ for which $f_1^8(\tau)-f_2^8(\tau)=af^2(\tau)$. This problem reduces to solving a quadratic equation (and calculation of the inverse of f). Then we compute $w_z(\tau)$ by formula (8.1), and compute $y_z(\tau)$ by formula (8.2). These are the roots of the equation considered.

We begin with studying the behavior of u and v under the changes of parameter τ (§§7.9–7.12). This is precisely the part of our program connected with the most cumbersome calculations; the theoretical background for these calculations is, however, quite simple.

§7.9. Transformations of order 5

To the matrix $\begin{pmatrix} a & b\\ c & d\end{pmatrix}$ we can assign a *fractional linear* or *Möbius* transformation

$$\tau\mapsto\frac{a\tau+b}{c\tau+d}.$$

Under this assignment the transformation $\tau\mapsto\tau+1$ corresponds to the matrix $T=\begin{pmatrix} 1 & 1\\ 0 & 1\end{pmatrix}$ and the transformation $\tau\mapsto-\frac{1}{\tau}$ corresponds to the matrix $S=\begin{pmatrix} 0 & -1\\ 1 & 0\end{pmatrix}$. It is easy to verify that the transformation $\tau\mapsto A(B\tau)$ corresponds to the matrix AB and to the matrix $E=\begin{pmatrix} -1 & 0\\ 0 & -1\end{pmatrix}$ there corresponds the identity transformation. Formulas (7.8) and (7.9) can be rewritten as follows:

$$\eta(T\tau)=e^{\frac{\pi i}{12}}\eta(\tau), \tag{9.1}$$

$$\eta(S\tau)=\sqrt{-i\tau}\,\eta(\tau). \tag{9.2}$$

Let $SL_2(\mathbb{Z})$ be the group of integer 2×2 matrices with determinant 1. One can show that the group $SL_2(\mathbb{Z})$ is generated by

$$T=\begin{pmatrix} 1 & 1\\ 0 & 1\end{pmatrix}\quad\text{and}\quad S=\begin{pmatrix} 0 & -1\\ 1 & 0\end{pmatrix}.$$

This means that if $A \in SL_2(\mathbb{Z})$, then $\eta(A\tau)$ can be expressed in terms of $\eta(\tau)$ with the help of formulas (9.1) and (9.2). This explicit expression in terms of the elements of the matrix A was first obtained by Dedekind and this is why the function $\eta(\tau)$ is called Dedekind's η-function.

One can also obtain a similar expression for $f(A\tau)$, but it involves not only $f(\tau)$ but also $f_1(\tau)$ or $f_2(\tau)$. For example,

$$f(T\tau) = e^{-\frac{\pi i}{24}} f_1(\tau).$$

We will only need these expressions for certain matrices A. We will prove these expressions separately and will not deduce a general formula for them.

The solution of the fifth degree equations is based on the study of transformations $f(\tau) \mapsto f(P\tau)$, where $P = \begin{pmatrix} p & q \\ r & s \end{pmatrix}$ is an integer matrix with determinant equal to 5. If $A \in SL_2(\mathbb{Z})$, then $f(AP\tau)$ can be expressed in terms of $f(P\tau)$ (or $f_i(P\tau)$ for $i = 1, 2$). Therefore, one has to find the simplest form to which P can be reduced by left multiplications by a matrix $A \in SL_2(\mathbb{Z})$.

Let c and d be relatively prime numbers such that $cp + dr = 0$. There exist integers a and b such that $ad - bc = 1$.

This means that

$$\begin{pmatrix} a & b \\ c & d \end{pmatrix} \in SL_2(\mathbb{Z}), \quad \begin{pmatrix} a & b \\ c & d \end{pmatrix}\begin{pmatrix} p & q \\ r & s \end{pmatrix} = \begin{pmatrix} p' & q' \\ 0 & s' \end{pmatrix},$$

where $p's' = 5$. Using the fact that

$$\begin{pmatrix} 1 & n \\ 0 & 1 \end{pmatrix}\begin{pmatrix} p & q \\ 0 & s \end{pmatrix} = \begin{pmatrix} p & q+ns \\ 0 & s \end{pmatrix},$$

we can reduce the matrix P to the form $\begin{pmatrix} p & q \\ 0 & s \end{pmatrix}$, where $-\frac{s}{2} \leq q \leq \frac{s}{2}$. As a result we see that P can be reduced to one of the following forms:

$$P_\infty = \begin{pmatrix} 5 & 0 \\ 0 & 1 \end{pmatrix}, \quad P_0 = \begin{pmatrix} 1 & 0 \\ 0 & 5 \end{pmatrix}, \quad P_{\pm 1} = \begin{pmatrix} 1 & \pm 1 \\ 0 & 5 \end{pmatrix}, \quad P_{\pm 2} = \begin{pmatrix} 1 & \pm 2 \\ 0 & 5 \end{pmatrix}.$$

Let $v_c = f(P_c\tau)$, i.e., $v_\infty(\tau) = f(5\tau)$ and $v_c(\tau) = f(\tau + c/5)$ for $c \neq \infty$. Let us try to find out how v_c behaves under the change of parameter $\tau \mapsto B\tau$. This change of parameter corresponds to the right multiplication of P_c by B. By multiplying P_cB from the left by a matrix $A \in SL_2(\mathbb{Z})$ we can obtain the matrix P_d, i.e., $f(P_cB\tau)$ can be expressed in terms of $v_d = f(P_d\tau)$ (or $f_i(P_d\tau)$).

The next three sections are devoted to the derivation of these expressions for the changes of parameter $\tau \mapsto \tau + 2$, $\tau \mapsto -\frac{1}{\tau}$ and $\tau \mapsto \frac{\tau-1}{\tau+1}$.

§7.10. The change of parameter $\tau \mapsto \tau + 2$

The change of parameter $\tau \mapsto \tau + 1$ sends $f(\tau)$ into $e^{-\frac{\pi i}{24}} f_1(\tau)$. Therefore, in order not to appeal to the function f_1, let us confine ourselves to the change of parameter $\tau \mapsto \tau + 2$. This change of parameter sends, for example, $f(5\tau)$ into $f(5\tau + 10) = e^{-\frac{5\pi i}{12}} f(5\tau)$ and $f(\frac{\tau}{5})$ into $f(\frac{\tau+2}{5})$.

These transformations look different: in one case we have a factor of $e^{-\frac{5\pi i}{12}}$, whereas there is no factor in the other case. The point is that our notations are ill adjusted. Selecting the functions v_c more adequately we can make all transformations look alike.

As was noted in §7.7, the period of the function f is equal to 48. Therefore, varying the integer c we only have 240 distinct functions $f(\frac{\tau+c}{5})$. Residues modulo 240 are divided into 48 five-tuples consisting of numbers with the same residue modulo 48. Select the numbers c from the same five-tuple by demanding that they be divisible by 48. This can be done by changing the notation:

$$v_c = f\left(\frac{\tau+c}{5}\right),$$

where $c \equiv 0 \pmod{48}$. We directly see that $v_c = v_d$ if $c \equiv d \pmod 5$. This means that for the index of v we may take the residue after the division of c by 5. Since $48 \equiv -2 \pmod 5$, it follows that

$$v_{\pm 2} = f\left(\frac{\tau \mp 48}{5}\right); \quad v_{\pm 1} = f\left(\frac{\tau \pm 96}{5}\right).$$

For v_0 and v_∞ we preserve the previous notation, i.e., $v_0 = f(\frac{\tau}{5})$ and $v_\infty = f(5\tau)$. This is the notation that we will use in the sequel. The change of parameter $\tau \mapsto \tau + 2$ replaces the function v_c with

$$f\left(\frac{\tau+2+c}{5}\right) = f\left(\frac{\tau+50-48+c}{5}\right) = e^{-\frac{5\pi i}{12}} f\left(\frac{\tau+c'}{5}\right),$$

where $c' = c - 48 \equiv 0 \pmod{48}$ and $c' \equiv c+2 \pmod 5$. Thus, v_c is replaced with $e^{-\frac{5\pi i}{12}} v_{c+2}$ and this also holds for $c = \infty$. Miraculously, the index c is transformed by the same rule as the parameter τ. In the next two sections we will see that this is also true for the changes of parameter $\tau \mapsto -\frac{1}{\tau}$ and $\tau \mapsto \frac{\tau-1}{\tau+1}$.

§7.11. The change of parameter $\tau \mapsto -\frac{1}{\tau}$

The change of parameter $\tau \mapsto -\frac{1}{\tau}$ replaces the function $v_\infty = f(5\tau)$ with

$$f\left(-\frac{\tau}{5}\right) = f\left(\frac{\tau}{5}\right) = v_0.$$

It is also easy to verify that v_0 is replaced with v_∞. Therefore, v_c is replaced with $v_{-1/c}$ for $c = 0, \infty$. For the other values of c this is also true but the proof requires cumbersome calculations.

To figure out the behavior of the functions v_c for $c \neq 0, \infty$, we have to represent the matrices from $SL_2(\mathbb{Z})$ in the form of the product of the matrices S and T. This is most easy to perform for matrices with small elements; therefore, we will use the fact that

$$v_{\pm 2} = f\left(\frac{\tau \mp 48}{5}\right) = f\left(\frac{\tau \pm 2}{5} \mp 10\right) = e^{\mp\frac{5\pi i}{12}} f\left(\frac{\tau \pm 2}{5}\right),$$

$$v_{\pm 1} = f\left(\frac{\tau \pm 96}{5}\right) = f\left(\frac{\tau \mp 4}{5} \pm 20\right) = e^{\mp\frac{5\pi i}{6}} f\left(\frac{\tau \mp 4}{5}\right).$$

The change of parameter $\tau \mapsto -\frac{1}{\tau}$ induces maps of functions

$$v_{\pm 2} \mapsto e^{\mp\frac{5\pi i}{12}} f\left(\frac{\pm 2\tau - 1}{5\tau}\right) \quad \text{and} \quad v_{\pm 1} \mapsto e^{\mp\frac{5\pi i}{6}} f\left(\frac{\mp 4\tau - 1}{5\tau}\right).$$

Let us start our calculations with the function $f(\frac{\tau+48}{5}) = v_{-2}$. For this function, we have to reduce the fractionally linear function $\frac{-2\tau-1}{5\tau}$ to one of the six main forms. Following the reduction algorithm described in §7.9 we get the relation

$$\begin{pmatrix} 2 & 1 \\ -5 & -2 \end{pmatrix} \begin{pmatrix} -2 & -1 \\ 5 & 0 \end{pmatrix} = \begin{pmatrix} 1 & -2 \\ 0 & 5 \end{pmatrix}.$$

Therefore,

$$f\left(\frac{-2\tau-1}{5\tau}\right) = f\left(A^{-1}\left(\frac{\tau-2}{5}\right)\right),$$

where

$$A^{-1} = \begin{pmatrix} 2 & 1 \\ -5 & -2 \end{pmatrix}^{-1} = -ST^2ST^{-2}S.$$

Hence,

$$f\left(\frac{-2\tau-1}{5\tau}\right) = f\left(ST^2ST^{-2}S\left(\frac{\tau-2}{5}\right)\right).$$

It is easy to verify that

$$f(ST^2ST^{-2}S\beta) = f(T^2ST^{-2}S\beta) = e^{-\frac{\pi i}{12}} f(ST^{-2}S\beta) = e^{-\frac{\pi i}{12}} e^{\frac{\pi i}{12}} f(\beta) = f(\beta).$$

Therefore, the function $f(\frac{\tau+48}{5})$ is replaced with

$$e^{-\frac{5\pi i}{12}} f\left(\frac{\tau-2}{5}\right) = e^{-\frac{5\pi i}{12}} f\left(\frac{\tau+48}{5} - 10\right) = f\left(\frac{\tau+48}{5}\right).$$

Similar calculations demonstrate that the function $f(\frac{\tau-48}{5})$ goes into $f(\frac{\tau-48}{5})$. Hence, v_c turns into v_c for $c = \pm 2$. It is also clear that $-\frac{1}{c} = c$ for $c = \pm 2$. For the function $f(\frac{\tau\pm 96}{5})$ calculations are somewhat lengthier. Since

$$\begin{pmatrix} 4 & 3 \\ 5 & 4 \end{pmatrix} \begin{pmatrix} -4 & -1 \\ 5 & 0 \end{pmatrix} = \begin{pmatrix} -1 & -4 \\ 0 & -5 \end{pmatrix}$$

and

$$\begin{pmatrix} 4 & 3 \\ 5 & 4 \end{pmatrix}^{-1} = STST^{-4}ST^{-1},$$

it follows that

$$f\left(\frac{-4\tau-1}{5\tau}\right) = f\left(STST^{-4}ST^{-1}\left(\frac{\tau+4}{5}\right)\right).$$

It is easy to verify that

$$\begin{aligned} f(STST^{-4}ST^{-1}\beta) &= f(TST^{-4}ST^{-1}\beta) = e^{-\frac{\pi i}{24}} f_1(ST^{-4}ST^{-1}\beta) \\ &= e^{-\frac{\pi i}{24}} f_2(T^{-4}ST^{-1}\beta) = e^{-\frac{9\pi i}{24}} f_2(ST^{-1}\beta) \\ &= e^{-\frac{9\pi i}{24}} f_1(T^{-1}\beta) = e^{-\frac{\pi i}{3}} f(\beta). \end{aligned}$$

Therefore, the function $f(\frac{\tau+96}{5})$ becomes $e^{-\frac{5\pi i}{6} - \frac{\pi i}{3}} f(\frac{\tau+4}{5})$. Clearly,

$$f\left(\frac{\tau+4}{5}\right) = f\left(\frac{\tau-96}{5} + 20\right) = e^{-\frac{5\pi i}{6}} f\left(\frac{\tau-96}{5}\right).$$

Since $\frac{5}{6} + \frac{1}{3} + \frac{5}{6} = 2$ and $e^{-2\pi i} = 1$, we see that $v_1 = f(\frac{\tau+96}{5})$ turns into $f(\frac{\tau-96}{5}) = v_{-1}$.

Similar calculations show that v_{-1} becomes v_1.

Thus, the change of parameter $\tau \mapsto -\frac{1}{\tau}$ sends the function v_c into $v_{-\frac{1}{c}}$ for all c.

§7.12. The change of parameter $\tau \mapsto \frac{\tau-1}{\tau+1}$

Let us prove that the change of parameter $\tau \mapsto \frac{\tau-1}{\tau+1}$ sends the function v_c into $-\frac{\sqrt{2}}{v_d}$, where $d = \frac{c-1}{c+1}$. In the proof of this statement the relation (7.16), i.e.,

$$f(\tau) f\left(\frac{\tau-1}{\tau+1}\right) = \sqrt{2},$$

plays an essential role. For the calculations we need a relation of the form

$$\begin{pmatrix} 1 & -1 \\ 1 & 1 \end{pmatrix} \begin{pmatrix} x & y \\ 0 & z \end{pmatrix} = \begin{pmatrix} \alpha & \beta \\ \gamma & \delta \end{pmatrix} \begin{pmatrix} a & b \\ 0 & d \end{pmatrix} \begin{pmatrix} 1 & -1 \\ 1 & 1 \end{pmatrix}, \tag{12.1}$$

where the matrix $\begin{pmatrix} a & b \\ 0 & d \end{pmatrix}$ is given and the matrix $\begin{pmatrix} \alpha & \beta \\ \gamma & \delta \end{pmatrix} \in SL_2(\mathbb{Z})$ is selected so that the lower left element of the matrix $\begin{pmatrix} x & y \\ 0 & z \end{pmatrix}$ is indeed 0. By multiplying (12.1) from the left by $\begin{pmatrix} 1 & 1 \\ -1 & 1 \end{pmatrix}$ we reduce this condition to the following one:

$$(-\alpha + \gamma)(a + b) + (\delta - \beta) d = 0.$$

The relations for v_0 and v_∞ are the easiest to obtain; they are of the form

$$\begin{pmatrix} 1 & -1 \\ 1 & 1 \end{pmatrix} \begin{pmatrix} 1 & -4 \\ 0 & 5 \end{pmatrix} = \begin{pmatrix} 1 & -4 \\ 0 & 1 \end{pmatrix} \begin{pmatrix} 5 & 0 \\ 0 & 1 \end{pmatrix} \begin{pmatrix} 1 & -1 \\ 1 & 1 \end{pmatrix},$$

$$\begin{pmatrix} 1 & -1 \\ 1 & 1 \end{pmatrix} \begin{pmatrix} 1 & 4 \\ 0 & 5 \end{pmatrix} = \begin{pmatrix} 1 & 0 \\ -4 & 1 \end{pmatrix} \begin{pmatrix} 1 & 0 \\ 0 & 5 \end{pmatrix} \begin{pmatrix} 1 & -1 \\ 1 & 1 \end{pmatrix}.$$

Therefore,

$$\frac{\sqrt{2}}{f(\frac{\tau-4}{5})} = e^{\frac{4\pi i}{24}} f\left(5\frac{\tau-1}{\tau+1}\right), \qquad \frac{\sqrt{2}}{f(\frac{\tau+4}{5})} = e^{-\frac{4\pi i}{24}} f\left(\frac{1}{5}\frac{\tau-1}{\tau+1}\right),$$

where in deducing the second formula we used the fact that

$$f\left(\frac{\tau}{-4\tau+1}\right) = f\left(\frac{4\tau-1}{\tau}\right) = e^{-\frac{4\pi i}{24}} f(\tau).$$

Taking into account that $e^{-\pi i} = -1$ we get

$$f\left(5\frac{\tau-1}{\tau+1}\right) = \frac{-\sqrt{2}}{f(\frac{\tau+96}{5})}, \qquad f\left(\frac{1}{5}\frac{\tau-1}{\tau+1}\right) = \frac{-\sqrt{2}}{f(\frac{\tau-96}{5})}.$$

The relations for v_1 and v_4 are of the form

$$\begin{pmatrix} 1 & -1 \\ 1 & 1 \end{pmatrix} \begin{pmatrix} 1 & 0 \\ 0 & 5 \end{pmatrix} = \begin{pmatrix} 3 & -1 \\ -2 & 1 \end{pmatrix} \begin{pmatrix} 1 & 1 \\ 0 & 5 \end{pmatrix} \begin{pmatrix} 1 & -1 \\ 1 & 1 \end{pmatrix}, \tag{12.2}$$

$$\begin{pmatrix} 1 & -1 \\ 1 & 1 \end{pmatrix} \begin{pmatrix} 5 & 0 \\ 0 & 1 \end{pmatrix} = \begin{pmatrix} 3 & 1 \\ 2 & 1 \end{pmatrix} \begin{pmatrix} 1 & -1 \\ 0 & 5 \end{pmatrix} \begin{pmatrix} 1 & -1 \\ 1 & 1 \end{pmatrix}. \tag{12.3}$$

Let us perform the calculations for v_1. Under the change of parameter $\tau \mapsto \frac{\tau-1}{\tau+1}$ the function $v_1 = f(\frac{\tau+96}{5})$ becomes

$$f\left(\frac{\alpha+96}{5}\right) = f\left(\frac{\alpha+1}{5}+19\right) = e^{-\frac{19\pi i}{24}} f_1\left(\frac{\alpha+1}{5}\right) = e^{-\frac{19\pi i}{24}} f_1(\beta),$$

where $\alpha = \frac{\tau-1}{\tau+1}$ and $\beta = \frac{\alpha+1}{5}$. Relation (12.2) means that

$$\frac{\sqrt{2}}{f(\frac{\tau}{5})} = f\left(\frac{3\beta-1}{-2\beta+1}\right).$$

It is easy to verify that

$$f\left(\frac{3\beta-1}{-2\beta+1}\right) = e^{\frac{\pi i}{24}} f_1\left(\frac{\beta}{-2\beta+1}\right) = e^{\frac{\pi i}{24}} f_2\left(2-\frac{1}{\beta}\right) = e^{\frac{5\pi i}{24}} f_1(\beta).$$

Hence,

$$e^{-\frac{19\pi i}{24}} f_1(\beta) = -e^{-\frac{5\pi i}{24}} f_1(\beta) = -\frac{\sqrt{2}}{f(\frac{\tau}{5})},$$

i.e., the function v_1 turns into $-\frac{\sqrt{2}}{f(\frac{\tau}{5})} = -\frac{\sqrt{2}}{v_0}$. Similar calculations demonstrate that v_4 becomes $-\frac{\sqrt{2}}{f(5\tau)} = -\frac{\sqrt{2}}{v_\infty}$.

For v_2 we get the relation

$$\begin{pmatrix} 1 & -1 \\ 1 & 1 \end{pmatrix}\begin{pmatrix} -1 & -2 \\ 0 & -5 \end{pmatrix} = \begin{pmatrix} -2 & 1 \\ 3 & -2 \end{pmatrix}\begin{pmatrix} 1 & 2 \\ 0 & 5 \end{pmatrix}\begin{pmatrix} 1 & -1 \\ 1 & 1 \end{pmatrix};$$

and then it is not difficult to verify that v_2 turns into $-\frac{\sqrt{2}}{v_2}$. We similarly prove that v_3 turns into $-\frac{\sqrt{2}}{v_3}$.

The results of the calculations performed in §§7.10–7.12 can be arranged in the following table, where $\varepsilon = e^{-\frac{5\pi i}{12}}$:

	u	v_∞	v_0	v_1	v_2	v_3	v_4
$\tau \mapsto \tau+2$	$e^{-\frac{\pi i}{12}}u$	εv_∞	εv_2	εv_3	εv_4	εv_0	εv_1
$\tau \mapsto \tau - \frac{1}{\tau}$	u	v_0	v_∞	v_4	v_2	v_3	v_1
$\tau \mapsto \frac{\tau-1}{\tau+1}$	$\frac{\sqrt{2}}{u}$	$-\frac{\sqrt{2}}{v_1}$	$-\frac{\sqrt{2}}{v_4}$	$-\frac{\sqrt{2}}{v_0}$	$-\frac{\sqrt{2}}{v_2}$	$-\frac{\sqrt{2}}{v_3}$	$-\frac{\sqrt{2}}{v_\infty}$

§7.13. Functions invariant with respect to the changes of parameter $\tau \mapsto \tau+2$, $\tau \mapsto -\frac{1}{\tau}$ and $\tau \mapsto \frac{\tau-1}{\tau+1}$

The function $f^{24}(\tau)$ is fixed under the changes of parameter $\tau \mapsto \tau+2$ and $\tau \mapsto -\frac{1}{\tau}$, whereas under the change of parameter $\tau \mapsto \frac{\tau-1}{\tau+1}$ it turns into $\frac{2^{12}}{f^{24}(\tau)}$.

If we consider the function

$$\begin{aligned} F(\tau) &= f^{24}(\tau) + \frac{2^{12}}{f^{24}(\tau)} \\ &= q^{-1}\prod\left(1+q^{2k-1}\right)^{24} + 2^{12}q\prod\left(1+q^{2k-1}\right)^{-24} = q^{-1}+24+\cdots, \end{aligned}$$

then, as is easy to verify, it does not change under the transformation of parameter $\tau \mapsto \frac{\tau-1}{\tau+1}$, as well. Indeed, this change of parameter turns $\frac{2^{12}}{f^{24}(\tau)}$ into $f^{24}(\tau)$.

We can prove that *if the function* $g(\tau)$ *is meromorphic in the upper half plane* $\operatorname{Im}\tau > 0$ *and does not change under the transformations* $\tau \mapsto \tau + 2$, $\tau \mapsto -\frac{1}{\tau}$ *and* $\tau \mapsto \frac{\tau-1}{\tau+1}$, *then, under a certain condition,* $g(\tau) = R(F(\tau))$, *where* R *is a rational function.* The condition in question is that the function $g(q)$, where $q = e^{\pi i\tau}$, should be meromorphic. For a meromorphic function $g(\tau)$ the condition reduces to the fact that the point $q = 0$ is not essentially singular, i.e., there are only finitely many nonzero coefficients c_n with negative n in the expansion $g(q) = \sum_{n=-\infty}^{\infty} c_n q^n$. (§§7.16–7.19 are devoted to the proof of this statement.)

Moreover, the equation $F(\tau) = c$ is solvable for all complex $c \neq 0$ (we will prove this in §7.19). If $R(F(\tau))$ is finite for all τ such that $F(\tau) \neq \infty$, then R is a polynomial. Indeed, let R be a fraction with a nonconstant denominator. Then its denominator vanishes for some $F(\tau) \neq \infty$. If $R(F)$ is finite for $q = 0$ as well (i.e., $F = \infty$), then R is a constant.

We will use these facts first to deduce the modular equation and then to solve quintic equations.

§7.14. Deduction of the modular equation

With the help of the transformation table (see §7.12) we can easily verify that the transformation law of the functions uv and u/v, where $u(\tau) = f(\tau)$, is as follows:

	uv	$\frac{u}{v}$
$\tau \mapsto \tau + 2$	$e^{-\frac{\pi i}{2}} uv$	$e^{\frac{\pi i}{3}} \frac{u}{v}$
$\tau \mapsto \tau - \frac{1}{\tau}$	uv	$\frac{u}{v}$
$\tau \mapsto \frac{\tau-1}{\tau+1}$	$-\frac{2}{uv}$	$-\frac{v}{u}$

Under these transformations the index c of the function v_c changes in the same way as τ.

Consider the functions $A_c = (u/v_c)^3 + (v_c/u)^3$ and $B_c = (uv_c)^2 - (4/(uv_c)^2)$. They transform as follows:

	A	B
$\tau \mapsto \tau + 2$	$-A$	$-B$
$\tau \mapsto -\frac{1}{\tau}$	A	B
$\tau \mapsto \frac{\tau-1}{\tau+1}$	$-A$	$-B$

As a result we see that the function $\prod_c (A_c - B_c)^2$, where the product is taken over $c = \infty, 0, 1, 2, 3, 4$, does not vary under all the above transformations of τ, since these transformations only permute factors.

It is easy to see that

$$A_\infty = q^{-\frac{1}{2}}(1 - 2q + \cdots), \quad B_\infty = q^{-\frac{1}{2}}(1 - 2q + \cdots).$$

Hence, $A_\infty - B_\infty$ vanishes at $q = 0$, i.e., as $\operatorname{Im}\tau \to \infty$. Let us demonstrate that $A_c - B_c$ vanishes at $q = 0$ for all c. Clearly,

$$u(5\tau - c) = f(5\tau) = v_\infty(\tau), \quad v_c(5\tau - c) = f\left(\frac{5\tau - c + c}{5}\right) = f(\tau) = u(\tau).$$

(Recall that in these formulas $c \equiv 0 \pmod{48}$.) Moreover, A and B do not vary under the interchange of u with v. Hence,

$$A_c(5\tau - c) = A_\infty(\tau), \quad B_c(5\tau - c) = B_\infty(\tau). \tag{14.1}$$

Since the conditions $\operatorname{Im}(5\tau - c) \to \infty$ and $\operatorname{Im}\tau \to \infty$ are equivalent, it follows that $A_c - B_c = 0$ vanishes at $q = 0$.

As a result, we see that the function $\prod_c (A_c - B_c)^2$ is a constant that vanishes at $q = 0$. Therefore, $A_c - B_c = 0$ for some c. Formulas (14.1) show that then $A_c - B_c = 0$ for all c. Therefore,

$$\left(\frac{u}{v}\right)^3 + \left(\frac{v}{u}\right)^3 = (uv)^2 - \frac{4}{(uv)^2},$$

i.e.,

$$v^6 - u^5v^5 + 4uv + u^6 = 0. \tag{14.2}$$

Equation (14.2) relates $u = f(\tau)$ with $v = f(5\tau)$; it is called the *modular equation*. Fixing $u = f(\tau)$ and considering (14.2) as an equation for v, we find that its roots are $v_\infty = f(5\tau)$ and $v_c = f(\frac{\tau+c}{5})$ for $c \equiv 0 \pmod{48}$. The *Viéta theorem*[2] implies, in particular, that

$$\prod v_c = u^6. \tag{14.3}$$

We will need this relation to solve the quintic equations.

§7.15. Solving quintic equations

In §7.14 we have shown that the coefficients of a sixth degree polynomial with roots v_c can be expressed in terms of u. Let us now prove that the coefficients of a fifth degree polynomial with roots w_0, w_1, w_2, w_3 and w_4, where

$$w_z = \frac{(v_\infty - v_z)(v_{z+1} - v_{z-1})(v_{z+2} - v_{z-2})}{\sqrt{5}u^3}, \tag{15.1}$$

can also be expressed in terms of u and find the explicit form of this polynomial.

With the help of the transformation table for v_c we can compose the following transformation table for w_z:

	w_0	w_1	w_2	w_3	w_4
$\tau \mapsto \tau + 2$	$-w_2$	$-w_3$	$-w_4$	$-w_0$	$-w_1$
$\tau \mapsto -\frac{1}{\tau}$	w_0	w_2	w_1	w_4	w_3
$\tau \mapsto \frac{\tau-1}{\tau+1}$	$-w_0$	$-w_3$	$-w_4$	$-w_2$	$-w_1$

[2] In Soviet mathematical schools the *Viéta theorem* was the statement that if x_1 and x_2 are the roots of the quadratic equation $x^2 + bx + c = 0$, then $x_1 + x_2 = -b$ and $x_1 x_2 = c$. *Translator.*

To calculate how w_z transforms under the change of parameter $\tau \mapsto \frac{\tau-1}{\tau+1}$, we must use relation (14.3).

Consider the polynomial

$$\prod(w - w_i) = w^5 + A_1w^4 + A_2w^3 + A_3w^2 + A_4w + A_5. \tag{15.2}$$

Its coefficients are finite at $u \neq 0, \infty$. Moreover, the functions A_1^2, A_2, A_3^2, A_4 and A_5^2 do not vary under the changes of parameter $\tau \mapsto \tau + 2$, $\tau \mapsto -\frac{1}{\tau}$ and $\tau \mapsto \frac{\tau-1}{\tau+1}$. Therefore, they are polynomials of $u^{24} + 2^{12}u^{-24} = q^{-1} + 24 + \cdots$. Such a polynomial is nonconstant only if its power series expansion in q begins with the term cq^r, where $r \leq -1$.

Let us calculate the first term in the expansion of w_z. Since

$$f(\tau) = q^{-\frac{1}{24}} \prod_{k=1}^{\infty}(1 + q^{2k-1}), \quad \text{where } q = e^{\pi i \tau},$$

the first term in the expansion of v_c is equal to $(q')^{-\frac{1}{24}}$, where $q' = e^{5\pi i\tau}$ for $c = \infty$ and $q' = e^{\frac{\pi i(\tau+c)}{5}}$ for $c \neq \infty$. It is easy to verify that $e^{-\frac{\pi i c}{24\cdot 5}} = \alpha^c$, where $\alpha = e^{-\frac{4\pi i}{5}}$. (Recall that $c \equiv 0 \pmod{48}$.) Therefore, the first term in the expansion of w_z is equal to

$$\frac{q^{-\frac{5}{24}}q^{-\frac{1}{120}}\left(\alpha^{z+1} - \alpha^{z-1}\right)q^{-\frac{1}{120}}\left(\alpha^{z+2} - \alpha^{z-2}\right)}{\sqrt{5}q^{-\frac{1}{8}}} = \lambda q^{-\frac{1}{10}},$$

where $\lambda = \frac{\alpha^{2z}\left(\alpha^3 - \alpha - \alpha^{-1} + \alpha^{-3}\right)}{\sqrt{5}} = \alpha^{2z}$, because

$$\alpha^3 - \alpha - \alpha^{-1} + \alpha^{-3} = 2\left(\cos 36^\circ + \cos 72^\circ\right) = \sqrt{5}.$$

Thus, the expansion of A_s begins with $q^{-\frac{s}{10}}$ (or with the term cq^r, where $r \geq -\frac{s}{10}$). Therefore, the functions A_1^2, A_2, A_3^2, and A_4 are constants whereas A_5^2 linearly depends on $u^{24}+2^{12}u^{-24} = q^{-1}+24+\cdots$ because its expansion begins with μq^{-1}, where $\mu = \left(\alpha^2\alpha^4\alpha^6\alpha^8\right)^2 = 1$. Comparing the first terms of the expansions of the functions $u^{24} + 2^{12}u^{-24}$ and A_5^2 we get

$$A_5^2 = u^{24} + \frac{2^{12}}{u^{24}} + C.$$

To calculate the constants C, A_1, A_2, A_3 and A_4, it suffices to calculate the value of v_c for one τ. Most convenient for calculations is $\tau = i$. Indeed, $-1/i = i$ and by (7.14) we get

$$f_1(i) = f_2(i). \tag{15.3}$$

It is also clear that for a purely imaginary τ the functions f, f_1 and f_2 assume positive values. Therefore, (15.3), (6.1) and (6.2) imply that $u = f(i) = \sqrt[4]{2}$. Using the fact that $(2 - i)(2 + i) = 5$ we get

$$\begin{aligned} v_3 &= f\left(\frac{i+48}{5}\right) = f\left(\frac{i-2}{5} + 10\right) = e^{-\frac{10\pi i}{24}} f\left(\frac{i-2}{5}\right) \\ &= e^{-\frac{10\pi i}{24}} f(i+2) = e^{-\frac{\pi i}{2}} f(i) = -i\sqrt[4]{2}. \end{aligned}$$

Similarly, $v_4 = i\sqrt[4]{2}$.

For $\tau = i$ the modular equation takes the form

$$v^6 - a^5v^5 + a^9v + a^6 = 0, \tag{15.4}$$

where $a = \sqrt[4]{2}$. We have already found two roots of this equation, namely, $v_3 = -ia$ and $v_2 = ia$. Dividing the polynomial (15.4) by $(v - v_2)(v - v_3) = v^2 + a^2$ we get

$$v^4 - a^5v^3 + a^2v^2 + a^7v + a^4 = (v - \alpha)^2(v - \beta)^2,$$

where $\alpha + \beta = a$ and $\alpha\beta = -a^2$. Assuming that $\alpha > 0$ and $\beta < 0$ we get $\alpha = \frac{a(1+\sqrt{5})}{2}$ and $\beta = \frac{a(1-\sqrt{5})}{2}$. It is also clear that

$$v_\infty = f(5i) = f(-1/5i) = f(i/5) = v_0$$

and we additionally have $v_\infty > 0$ since $f(\tau) > 0$ for any purely imaginary τ. Therefore,

$$v_0 = v_\infty = \frac{\sqrt[4]{2}\,(1 + \sqrt{5})}{2}, \quad v_1 = v_4 = \frac{\sqrt[4]{2}\,(1 - \sqrt{5})}{2}.$$

Substituting the value of v_c at $\tau = i$ into (15.1) we get

$$w_0 = 0, \quad w_1 = w_2 = i\sqrt{5}, \quad w_3 = w_4 = -i\sqrt{5}.$$

Therefore, for $\tau = i$ the polynomial (15.2) is of the form

$$w(w - i\sqrt{5})^2(w + i\sqrt{5})^2 = w(w^2 + 5)^2.$$

Thus, $A_5^2(i) = 0$, i.e.,

$$C = -\left(u^{24}(i) + \frac{2^{12}}{u^{24}(i)}\right) = -\left(2^6 + 2^6\right) = -2^7.$$

Hence,

$$A_5^2 = u^{24} + \frac{2^{12}}{u^{24}} - 2^7 = \left(u^{12} - \frac{2^6}{u^{12}}\right)^2,$$

i.e.,

$$A_5 = \pm\left(u^{12} - \frac{2^6}{u^{12}}\right).$$

It is easy to verify that the expansions of functions $A_5 = -w_0w_1w_2w_3w_4$ and $u^{12} - 2^6u^{-12}$ begin with $-q^{\frac{1}{2}}$ and $q^{\frac{1}{2}}$, respectively. Hence, $A_5 = -u^{12} + 2^6u^{-12}$ and, therefore, equation (15.2) is of the form

$$w\left(w^2 + 5\right)^2 = u^{12} - 64u^{-12}.$$

The relations $f^8 = f_1^8 + f_2^8$ and $ff_1f_2 = \sqrt{2}$ imply that

$$u^{12} - \frac{64}{u^{12}} = \frac{f^{24}(\tau) - 64}{f^{12}(\tau)} = \left(\frac{f_1^8(\tau) - f_2^8(\tau)}{f^2(\tau)}\right)^2.$$

Hence,

$$\sqrt{w(\tau)} = \pm\frac{f_1^8(\tau) - f_2^8(\tau)}{f^2(\tau)\,(w^2(\tau) + 5)}.$$

Set

$$y(\tau) = \frac{f_1^8(\tau) - f_2^8(\tau)}{f^2(\tau)\,(w^2(\tau) + 5)}. \tag{15.5}$$

Then

$$y^5+5y=y\left(w^2+5\right)=\frac{f_1^8-f_2^8}{f^2}.$$

If $\left(f_1^8-f_2^8\right)/f^2=a$, then the roots of the equation

(15.6) $$y^5+5y=a$$

can be calculated as follows. First, calculate $v_c(\tau)$ for $c=\infty$, 0, 1, 2, 3, 4. Next with the help of formula (15.1) calculate $w_z(\tau)$ for $z=0$, 1, 2, 3, 4. Finally, calculate $y_z(\tau)$ for $z=0$, 1, 2, 3, 4 by formula (15.5). These are the roots of equation (15.6). To learn how to solve equation (15.6) for any value of the parameter a, we only have to learn how to solve equation

(15.7) $$f_1^8(\tau)-f_2^8(\tau)=af^2(\tau).$$

Taking into account that $f_1^8+f_2^8=f^8$ and $ff_1f_2=\sqrt{2}$ and squaring (15.7) we reduce it to the form

(15.8) $$f^{24}-a^2f^{12}-64=0.$$

Equation (15.8) is a quadratic equation for f^{12}. One of its roots gives a solution of equation (15.7), the other root gives a solution of the equation obtained from (15.7) under the change of parameter $a\mapsto -a$.

* * *

In §7.13 we have formulated certain properties of functions invariant under the changes of parameter $\tau\mapsto\tau+2$, $\tau\mapsto-\frac{1}{\tau}$ and $\tau\mapsto\frac{\tau-1}{\tau+1}$. We used these properties to solve quintic equations. Let us now prove these properties.

§7.16. The main modular function $j(\tau)$

In this section we will construct a function $j(\tau)$ invariant under the changes of parameter $\tau\mapsto\tau+1$ and $\tau\mapsto-\frac{1}{\tau}$. The function $j(\tau)$ is of interest thanks to the fact that any function invariant under these changes and satisfying certain meromorphicity conditions can be represented as a rational function of j.

First, consider the functions

$$k(\tau)=\frac{\Theta_2^2(0|\tau)}{\Theta_3^2(0|\tau)},\quad k'(\tau)=\frac{\Theta_0^2(0|\tau)}{\Theta_3^2(0|\tau)}.$$

They are related by equation $k^2+k'^2=1$ (cf. §7.3). The transformation formulas of the theta functions with respect to the parameter τ obtained in §7.7 enable us to verify that

$$k(\tau+1)=i\frac{k(\tau)}{k'(\tau)},\quad k'(\tau+1)=\frac{1}{k'(\tau)};$$
$$k(-1/\tau)=k'(\tau),\quad k'(-1/\tau)=k(\tau).$$

Now, consider the function $\lambda(\tau)=\left(k'(\tau)\right)^2$. The changes $\tau\mapsto\tau+1$ and $\tau\mapsto-\frac{1}{\tau}$ send λ into λ^{-1} and $1-\lambda$, respectively. We want to obtain a function $j(\tau)$ invariant under these changes of parameter. To this end it suffices to construct a rational function of λ invariant under the interchanges of λ with λ^{-1} and $1-\lambda$.

Such a function can be obtained as follows. Under the action of the above transformations, the number λ can become one of the following numbers:

$$\lambda, \quad \frac{1}{\lambda}, \quad 1-\lambda, \quad \frac{1}{1-\lambda}, \quad \frac{\lambda}{1-\lambda}, \quad \frac{\lambda-1}{\lambda}.$$

Therefore, the function

$$J_1(\lambda) = \lambda^2 + \frac{1}{\lambda^2} + (1-\lambda)^2 + \frac{1}{(1-\lambda)^2} + \left(\frac{\lambda}{1-\lambda}\right)^2 + \left(\frac{\lambda-1}{\lambda}\right)^2$$

is invariant under the transformations indicated.

The two other functions obtained from J_1 by linear transformations, namely,

$$J_2 = \frac{J_1+3}{2} = \frac{(\lambda^2-\lambda+1)^3}{\lambda^2(1-\lambda)^2}$$

and

$$J_3 = J_2 - \frac{27}{4} = \left(\frac{(\lambda+1)(\lambda-2)(\lambda-\frac{1}{2})}{\lambda(1-\lambda)}\right)^2,$$

are often more convenient. The function J_2 is, actually, the most convenient one for our purposes. Set

$$\begin{aligned} j(\tau) &= 2^8 J_2 = 2^8 \frac{(1-\lambda+\lambda^2)^3}{\lambda^2(1-\lambda)^2} = 2^8 \frac{(1-k^2k'^2)^3}{k^4k'^4} \\ &= 2^8 \frac{(\Theta_3^8 - \Theta_2^4\Theta_0^4)^3}{\Theta_0^8\Theta_2^8\Theta_3^8} = (2\pi)^8 \frac{(\Theta_3^8 - \Theta_2^4\Theta_0^4)^3}{\Theta_1'^8} = (f^{16} - f_1^8 f_2^8)^3 = \frac{(f^{24}-16)^3}{f^{24}}. \end{aligned}$$

Using the fact that $f^{24}(\tau) = q^{-1}\prod(1+q^{2k-1})^{24}$ we can obtain the following expansion for j:

$$\begin{aligned} j(\tau) &= \left(f^{16} - \frac{16}{f^8}\right)^3 = q^{-2}\prod(1+q^{2k-1})^{48} - 48q^{-1}\prod(1+q^{2k-1})^{24} \\ &\quad + 3\cdot 16^2 - 16^3 q \prod(1+q^{2k-1})^{-24} = q^{-2} + 744 + \sum_{n=1}^{\infty} c_n q^n. \end{aligned}$$

The function $j(\tau)$ is invariant under the changes of variable $\tau \mapsto \tau+1$ and $\tau \mapsto -\frac{1}{\tau}$. Since the change of variable $\tau \mapsto \tau+1$ sends $q = e^{\pi i\tau}$ into $-q$, the function $j(q)$ is even. Hence, all the coefficients c_n with odd indices vanish.

§7.17. The fundamental domain of $j(\tau)$

The function $j(\tau)$ is defined on the upper half plane $H = \{\tau \in \mathbb{C} \mid \operatorname{Im}\tau > 0\}$. At the points obtained from each other under the transformations $\tau \to \tau \pm 1$ and $\tau \mapsto -\frac{1}{\tau}$ the values of j coincide.

Let us show that the set $D = \{\tau \in H \mid |\tau| \geq 1, |\operatorname{Re}\tau| \leq 1/2\}$ possesses the following two properties:

1) Any point $\tau \in H$ can be sent into a point $\tau' \in D$ by a composition of transformations $\tau \mapsto \tau \pm 1$ and $\tau \mapsto -\frac{1}{\tau}$.

2) No two distinct inner points of D can be transformed into each other by a transformation indicated in property 1).

Let $G = SL_2(\mathbb{Z})/\{\pm I\}$ be the group of fractional linear transformations of the form

$$\tau \mapsto \frac{a\tau + b}{c\tau + d}, \quad \text{where} \quad \begin{pmatrix} a & b \\ c & d \end{pmatrix} \in SL_2(\mathbb{Z}),$$

i.e., $a, b, c, d \in \mathbb{Z}$ and $ad - bc = 1$.

The transformations $S : \tau \mapsto \frac{-1}{\tau}$ and $T : \tau \mapsto \tau + 1$ belong to G. Therefore the group G' generated by S and T is a subgroup of G. This remark enables us to prove that for a fixed $\tau \in H$ one can select among all points $g'\tau$, $g' \in G'$, the one with the maximal imaginary part (and it is impossible to select one with the minimal imaginary part since $\lim_{n\to\infty} \operatorname{Im}\left(-\frac{1}{\tau+n}\right) = 0$). If $g \in G$, then

$$\operatorname{Im}(g\tau) = \operatorname{Im}\left(\frac{a\tau + b}{c\tau + d}\right) = \operatorname{Im}\frac{ad\tau + bc\bar{\tau}}{|c\tau + d|^2} = \frac{\operatorname{Im}\tau}{|c\tau + d|^2}.$$

Therefore, the inequality $\operatorname{Im}(g\tau) \geq \operatorname{Im}(\tau)$ is equivalent to the inequality $|c\tau + d| \leq 1$ which is only true for finitely many pairs of integers (c, d). Thus, $\operatorname{Im}(g\tau)$ takes finitely many values larger than or equal to $\operatorname{Im}(\tau)$ (although each such value is taken for an infinite set of elements $g \in G$). As a result we see that in the search of an element $g \in G$ for which the imaginary part of $g\tau$ is maximal we can confine ourselves to a finite set of elements. These arguments hold not only for the whole group G but for any subgroup of G as well.

Let τ' have the maximal imaginary part among the images of $\tau \in H$ under the G'-action. Since the transformations $\tau' \mapsto \tau' \pm 1$ do not change the imaginary part of τ', we may assume that $|\operatorname{Re}\tau'| \leq \frac{1}{2}$. Let us show that in this case $\tau' \in D$, i.e., $|\tau'| \leq 1$. Indeed, the condition $\operatorname{Im}\tau' \geq \operatorname{Im}(g'\tau')$ implies, in particular, that

$$\operatorname{Im}\tau' \geq \operatorname{Im}\left(-\frac{1}{\tau'}\right) = \frac{1}{|\tau'|^2}\operatorname{Im}\tau',$$

i.e., $|\tau'| \geq 1$.

Now we prove a refined version of property 2). Namely, we show that under the action of the elements of G' the boundary points of the domain D are identified as shown on Figure 42, i.e., the rays QP and SP are glued (under the transformations

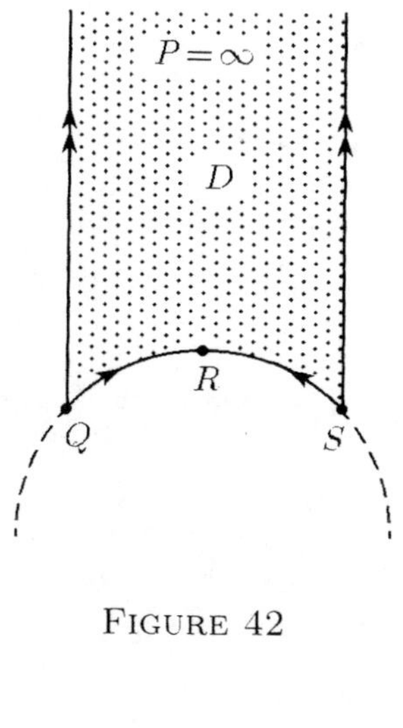

FIGURE 42

$\tau \mapsto \tau \pm 1$), the arcs $\smile QR$ and $\smile SR$ are also glued (under the transformation $\tau \mapsto -\frac{1}{\tau}$). No other boundary points are glued. Observe that if the infinite point P belongs to D, then after the indicated gluing D becomes a sphere, i.e., j can be considered as a function on a sphere.

In the following section we will show that j defines a one-to-one map of this sphere onto $\mathbb{C} \cup \infty$.

Suppose that $\tau' = g'\tau$, where $g' \in G'$, and let τ and τ' be two distinct points of D. We may assume that

$$g'\tau = \frac{a\tau + b}{c\tau + d}, \quad \begin{pmatrix} a & b \\ c & d \end{pmatrix} \in SL_2(\mathbb{Z}).$$

First, let us consider the case $c = 0$. In this case $ad = 1$, i.e., the transformation is of the form $\tau \mapsto \tau \pm b$. For case $b = 0$ the points τ and τ' coincide. If $b \neq 0$, then the image of D intersects with D only for transformations $\tau \mapsto \tau \pm 1$, and their intersection belongs to the set $|\mathrm{Re}\,\tau| = \frac{1}{2}$.

Now suppose that $c \neq 0$. Then

$$\tau' = \frac{a\tau + b}{c\tau + d} = \frac{a}{c} - \frac{1}{c(c\tau + d)},$$

i.e.,

$$(17.1) \qquad \left|\tau' - \frac{a}{c}\right| \cdot \left|\tau + \frac{d}{c}\right| = \frac{1}{c^2}.$$

The numbers $\frac{a}{c}$ and $\frac{d}{c}$ are real, and, therefore, the imaginary parts of the numbers $\tau' - \frac{a}{c}$ and $\tau + \frac{d}{c}$ are equal to the imaginary parts of the numbers τ' and τ, and, therefore, the absolute values of τ' and τ are not less than $\frac{\sqrt{3}}{2}$. Hence, $|c| \leq \frac{2}{\sqrt{3}}$. Since c is a nonzero integer, we have $c = \pm 1$. Therefore, (17.1) can be expressed in the form $|\tau' \mp a| \cdot |\tau \pm d| = 1$. If $\tau \in D$ and $m \in \mathbb{Z}$, then $|\tau + m| \geq 1$, where the equality takes place in the following cases only:

1) $|\tau| = 1, m = 0$;
2) $\tau = \exp(\pi i/3)$, $m = 0, -1$;
3) $\tau = \exp(2\pi i/3)$, $m = 0, 1$;

Therefore, if $c \neq 0$, then only points whose absolute value is equal to 1 can be glued; moreover, to glue to the points other than $\exp(k\pi i/3)$ is only possible via the map $\tau \mapsto -\frac{1}{\tau}$. It is also possible to glue to the points $\exp(k\pi i/3)$ via the maps $\tau \mapsto -\frac{1}{\tau} \pm 1$.

Let us notice that we have proved one more refinement of property 2). Namely, *if τ is an inner point of D and $g\tau \in D$, where $g \in G$, then g is the identity transformation.* With the help of this property we can prove the following statement.

7.17.1. THEOREM. *The group $G = SL_2(\mathbb{Z})/\{\pm I\}$ is generated by elements S and T, i.e., $G' = G$.*

PROOF. Let g be an arbitrary element of G. Consider an inner point τ_0 of D. Since

$$\mathrm{Im}(g\tau) = \frac{\mathrm{Im}\,\tau}{|c\tau + d|^2},$$

it follows that $g\tau_0 \in H$. Therefore, there exists an element g' of G' such that $g'(g\tau_0) \in D$. As a result we see that the inner point τ_0 of D turns into a point of the domain D under the transformation $g'g \in G$. Hence, $g'g$ is the identity transformation, so that $g = (g')^{-1} \in G'$. □

If we wish the domain D to have no distinct points obtained from each other under transformations from G, we can achieve this, for instance, by excluding the ray PQ and the arc $\smile QR$ from D (the points P, Q and R themselves are not excluded). The obtained set is called the *fundamental domain* of the group G. If we do not distinguish the points obtained from each other under the transformations S and T, then the domain D itself can be called the fundamental domain. The fundamental domain can be endowed with the structure of a topological space, since it is the quotient of H modulo the G-action.

The value of j at any point $\tau \in H$ coincides with the value of j at the corresponding point $\tau' \in D$. Moreover, as we will show in the next section the values of j at distinct points of the fundamental domain of G are distinct. This is why D is also called the *fundamental domain of the function* j.

§7.18. How to solve the equation $j(\tau) = c$

The function $j(\tau) = q^{-2} + 744 + \cdots$ has no singularities in the upper half plane $H = \{t \mid \operatorname{Im} \tau > 0\}$. Hence, the function j has no poles in the upper half plane H. For $\operatorname{Im} \tau = \infty$ the function $j(\tau)$ is infinite.

We start solving the equation $j(\tau) = c$ by computing the values of j at the "vertices" of the fundamental domain, i.e., at the points $i, \varepsilon = e^{\frac{\pi i}{3}}$ and ε^2. Recall that $j(\tau) = \left(f^{16} - \frac{16}{f^8}\right)^3$ and $f(i) = \sqrt[4]{2}$ (cf. §7.12). Hence, $j(i) = 12^3 = 1728$. To calculate $j(\varepsilon) = j(\varepsilon^2)$, we can make use of the fact that $\varepsilon = 1 - \frac{1}{\varepsilon}$. Indeed, this relation immediately implies:

$$f(\varepsilon) = f\left(1 - \frac{1}{\varepsilon}\right) = e^{-\frac{\pi i}{24}} f_2(\varepsilon), \quad f_1(\varepsilon) = f_1\left(1 - \frac{1}{\varepsilon}\right) = e^{-\frac{\pi i}{24}} f(\varepsilon).$$

Hence, $\sqrt{2} = f(\varepsilon) f_1(\varepsilon) f_2(\varepsilon) = f^3(\varepsilon)$ and, therefore, $f^{24}(\varepsilon) = \left(\sqrt{2}\right)^8 = 16$. Hence, $j(\varepsilon) = 0$.

7.18.1. THEOREM. *The function $j(\tau)$ takes every value c in the fundamental domain exactly once.*

PROOF. First suppose that c is distinct from 0 and 1728. If the value c is taken on the boundary of the fundamental domain, we can modify the fundamental domain as indicated in Figure 43 (on the next page). More exactly, if the value c is attained at a boundary point, then we cut a neighborhood U of this point and attach to D the image of U under one of the maps S or $T^{\pm 1}$. Such an operation is impossible to perform for the "vertices" but we cannot encounter them since $c \neq 0, 1728$.

If the meromorphic function $g(z)$ has no poles in G and no zeros on the boundary ∂G, then the number of zeros of g inside G is equal to

$$\frac{1}{2\pi i} \int_{\partial G} \frac{g'(z)}{g(z)} dz.$$

Indeed, let $g(z) = c_0(z - a)^r + c_1(z - a)^{r+1} + \cdots$. Then

$$g'(z)/g(z) = r(z - a)^{-1} + \cdots.$$

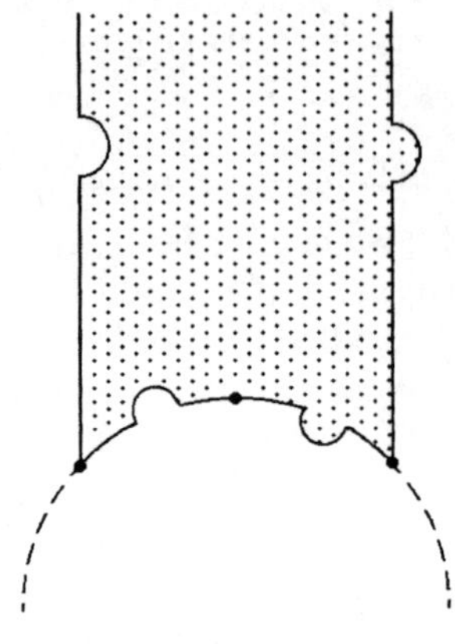

FIGURE 43

Therefore, the sum of residues at the singular points of g'/g belonging to G is equal to the number of zeros of g (multiplicities counted).

For G take the domain plotted in Figure 44, and for g take the function $j(z)-c$.

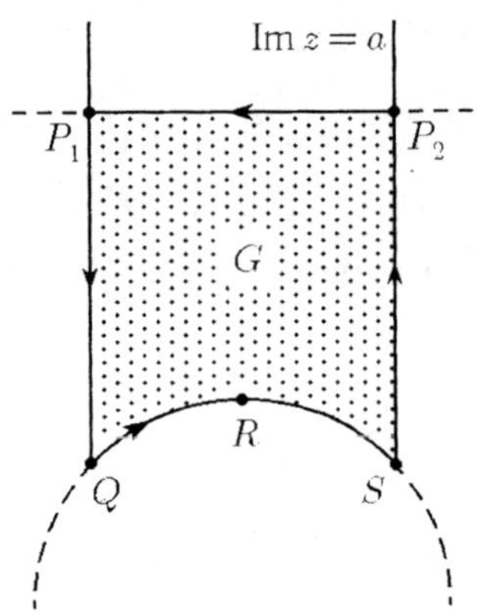

FIGURE 44

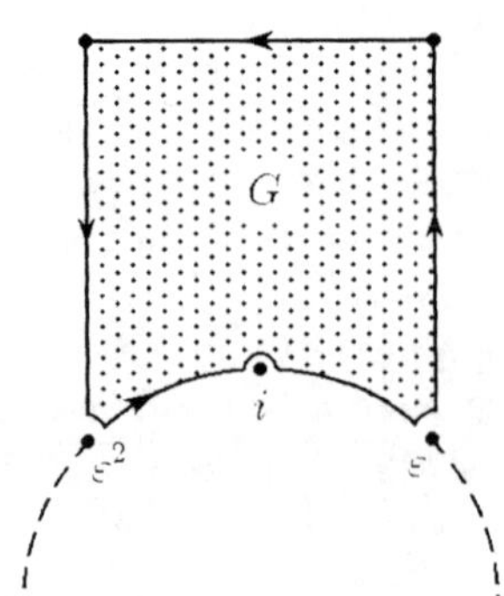

FIGURE 45

It is easy to verify that

$$\int_{\partial G} \frac{j'(z)dz}{j(z)-c} = \int_{P_1}^{P_2} \frac{j'(z)dz}{j(z)-c} = \int_{ia+\frac{1}{2}}^{ia-\frac{1}{2}} d\ln(j(z)-c).$$

Indeed, $j(z+1) = j(z)$ and, therefore, the integrals along the line segments P_1Q and SP_2 cancel. Moreover, $j(-1/z) = j(z)$ and the integrals along the arcs $\smile QR$ and $\smile RS$ also cancel. (This statement remains true even if we have to modify the boundary of the fundamental domain as well.)

Since $j(\tau) = q^{-2} + 744 + \cdots = e^{-2\pi i\tau} + 744 + \cdots$ for sufficiently large a and $z \in \left[ia - \frac{1}{2}, ia + \frac{1}{2}\right]$, we have

$$d\ln(j(z)-c) = d(-2\pi iz + \cdots) = -2\pi i dz + \cdots .$$

Therefore,

$$\lim_{a\to\infty} \frac{1}{2\pi i} \int_{ia+\frac{1}{2}}^{ia-\frac{1}{2}} d\ln(j(z)-c) = 1.$$

If $c = 0$ or 1728, we can proceed as follows. Consider the domain G plotted in Figure 45. (If the value c is attained on the boundary, then we have to modify the domain as before.)

The number of zeros of the function $j(z) - c$ belonging to G is equal to

$$\frac{1}{2\pi i} \int_{\partial G} \frac{j'(z)dz}{j(z)-c} = 1 - \frac{r(i)}{2} - \frac{r(\varepsilon)}{3},$$

where $r(i)$ and $r(\varepsilon)$ are the multiplicities of the zeros of the function $j(z) - c$ at the points i and ε, respectively. To prove this statement, it suffices to observe that for i we take the integral over the half circle, whereas for the points ε and ε^2 the integral is taken along a third of this circle and the circle is circumvented clockwise which leads to the minus sign.

If $c = 0$ or 1728, then either $r(\varepsilon) \neq 0$ or $r(i) \neq 0$. In this case the number of zeros of $j(\tau) - c$ belonging to G is strictly less than 1, i.e., $j(\tau) \neq c$ for $\tau \in G$. Observe also that if $c = 0$, then $r(\varepsilon) = 3$, and if $c = 1728$, then $r(i) = 2$. This means that the function j has a zero of multiplicity 3 at ε, whereas $j - 1728$ has a zero of multiplicity 2 at i. □

§7.19. The functions invariant under the changes of parameter $\tau \mapsto \tau + 1$ and $\tau \mapsto -\frac{1}{\tau}$

Any function $g(\tau)$ defined in the upper half plane H and invariant under the changes of variables $\tau \mapsto \tau+1$ and $\tau \mapsto -\frac{1}{\tau}$ can be represented in the form $G(j(\tau))$ for some function G. Indeed, as was shown in the preceding section, j determines a one-to-one map of $D \cup \infty$ onto $\mathbb{C} \cup \infty$. Therefore, there exists an inverse map $j^{-1} : \mathbb{C} \cup \infty \mapsto D \cup \infty$. Set $G(z) = g(j^{-1}(z))$. Then $G(j(\tau)) = g(j^{-1}(j(\tau))) = g(\tau)$.

Now suppose that the function g considered as a function of $q = e^{\pi i\tau}$ is meromorphic in the disk $|q| \leq 1$. Then G is a rational function. Indeed, since g is meromorphic, the finite singular points of G can only be poles. The value $j = \infty$ corresponds to $q = 0$; hence, by the assumption on g the point ∞ cannot be an essentially singular point for G. Therefore, in $\mathbb{C} \cup \infty$ the function G does not have singular points other than the poles. Let us subtract from G the principal parts of its power series expansions at the singular points. As a result we get a function without singular points in $\mathbb{C} \cup \infty$, i.e., a constant. Since all the principal parts of the expansions of G at singular points consist of finitely many terms, we deduce that G is a rational function.

§7.20. The functions invariant with respect to the changes of parameter $\tau \mapsto \tau + 2$ and $\tau \mapsto -\frac{1}{\tau}$

To solve a quintic equation we needed a description of the functions invariant with respect to changes of parameter $\tau \mapsto \tau + 2$, $\tau \mapsto -\frac{1}{\tau}$ and $\tau \mapsto \frac{\tau-1}{\tau+1}$. So far, we have only described the functions invariant with respect to the changes of parameter $\tau \mapsto \tau + 1$ and $\tau \mapsto -\frac{1}{\tau}$. Now it is easy to get a description of the functions we are interested in.

Recall that

$$j(\tau) = \frac{(f^{24}(\tau) - 16)^3}{f^{24}(\tau)} = \left(\frac{f^{24}(\tau) - 16}{f^8(\tau)}\right)^3 = \left(f^{16}(\tau) - \frac{16}{f^8(\tau)}\right)^3.$$

The function $f^{24}(\tau)$ is invariant with respect to the changes of parameter $\tau \mapsto \tau+2$ and $\tau \mapsto -\frac{1}{\tau}$. In the class of functions stable under these changes it plays the same role as the function $j(\tau)$ plays in the class of functions stable under the changes of parameter $\tau \mapsto \tau + 1$ and $\tau \mapsto -\frac{1}{\tau}$. Namely, any function g invariant under the changes of parameter $\tau \mapsto \tau + 2$ and $\tau \mapsto -\frac{1}{\tau}$ is a rational function of f^{24} (provided the function g considered as a function of $q = e^{i\pi\tau}$ is meromorphic in the disk $|q| < 1$). This can be proved by the same method that we used in §7.19. We have only to prove the following statement.

7.20.1. THEOREM. a) *The fundamental domain of the group G_1 generated by the transformations $\tau \mapsto \tau + 2$ and $\tau \mapsto -\frac{1}{\tau}$ is*

$$D_1 = \{\tau \in H \mid |\tau| \geq 1, \ |\operatorname{Re}\tau| \leq 1\}.$$

b) *In the domain D_1 the function f^{24} attains each nonzero value c exactly once.*

PROOF. Let τ' be the image of $\tau \in H$ under the action of G_1 with the maximal value of $\operatorname{Im}\tau'$ (the existence of such a point τ' is proved in §7.17). Since the transformation $\tau' \mapsto \tau' \pm 2$ does not change the imaginary part of τ', we can assume that $|\operatorname{Re}\tau'| \leq 1$. In this case $\tau' \in D_1$, i.e., $|\tau'| \geq 1$. Indeed, since $\operatorname{Im}\tau'$ is maximal, it follows that

$$\operatorname{Im}\tau' \geq \operatorname{Im}\left(-\frac{1'}{\tau}\right) = \operatorname{Im}\tau'/|\tau'|^2;$$

hence, $|\tau'| \geq 1$.

We proved that any point $\tau \in H$ can be mapped to a point $\tau' \in D_1$ under the action of some $g \in G_1$. To prove that distinct inner points of D_1 cannot be transformed into each other under the action of elements from G_1 it suffices to verify b). Let us do this.

It is convenient to complete the domain D_1 (Figure 46) with the points ± 1 (corresponding to the value $q = -1$). For these points q we get

$$f^{24} = q^{-1} \prod_{k=1}^{\infty} \left(1 + q^{2k-1}\right)^{24} = 0$$

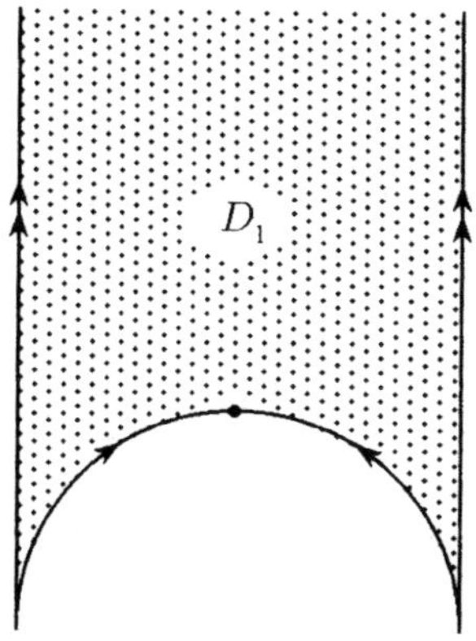

FIGURE 46

and this zero is of infinite multiplicity, i.e., for the function $j = \left(f^{16} - 16/f^8\right)^3$ the points $\tau = \pm 1$ are essentially singularities. In the domain D_1 the function j has no poles, therefore, the function f^{24} does not vanish anywhere in D_1. Moreover, for $\operatorname{Im}\tau = \infty$ the values of the functions j and f^{24} are equal to ∞. It remains to consider the nonzero values of f^{24}.

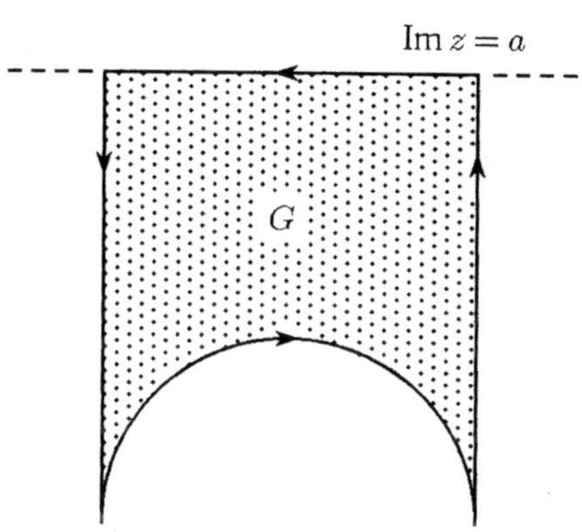

FIGURE 47

Let G be the domain plotted in Figure 47. Taking into account the gluings shown in Figure 46 we get for $c \neq 0$

$$\int_{\partial G} d\ln(f^{24}(z) - c) = \int_{ia+\frac{1}{2}}^{ia-\frac{1}{2}} d\ln(f^{24}(z) - c).$$

Since $f^{24}(\tau) = q^{-1} + \cdots = e^{-\pi i\tau} + \cdots$, it follows that

$$d\ln(f^{24}(z) - c) = d(-\pi iz + \cdots) = -\pi idz + \cdots .$$

Hence,

$$\lim_{a\to\infty} \frac{1}{2\pi i} \int_{ia+\frac{1}{2}}^{ia-\frac{1}{2}} d\ln(f^{24}(z) - c) = 1.$$

This means that in D_1 the function f^{24} assumes each nonzero value exactly once. □

Now, suppose that the function g satisfying the meromorphicity condition is invariant not only with respect to the changes of parameter $\tau \mapsto \tau + 2$ and $\tau \mapsto -\frac{1}{\tau}$ but also with respect to the change $\tau \mapsto \frac{\tau-1}{\tau+1}$. Then $g(\tau) = R(f^{24}(\tau))$, where R is a rational function such that $R(f^{24}(\frac{\tau-1}{\tau+1})) = R(f^{24}(\tau))$, i.e., $R(\frac{2^{12}}{f^{24}}) = R(f^{24})$.

Set $f^{24}(\tau) = x$ and $2^{12} = a$, and let $R(x) = \sum_{k=-\infty}^{\infty} c_k x^k$. Then

$$R\left(\frac{a}{x}\right) = \sum_{k=-\infty}^{\infty} c_k a^k x^{-k}.$$

Hence, $c_k a^k = c_{-k}$, so that

$$R(x) = c_0 + \sum_{k=1}^{\infty} c_k \left(x^k + \left(\frac{a}{x}\right)^k\right).$$

Considering the differences $\left(x + \frac{a}{x}\right)^k - \left(x^k + \left(\frac{a}{x}\right)^k\right)$ it is easy to prove by induction on k that $x^k + \left(\frac{a}{x}\right)^k$ can be polynomially expressed in terms of $x + \frac{a}{x}$. Therefore, $R(x) = G(x + \frac{a}{x})$ for some function G. Since R is rational, G has no singular points other than poles, i.e., G is a rational function. As a result, we see that the function $g(\tau)$ can be rationally expressed in terms of the function $f^{24}(\tau) + \frac{2^{12}}{f^{24}(\tau)}$.

Bibliography

A. Classical works

[A1] N. H. Abel, *Oeuvres complètes du Niels Henrik Abel*, Christiania, 1881.

[A2] J. Bernoulli, *Opera Omnia, T. I–IV*, Bosquet, Lausanne, Genevae, 1742.

[A3] Diophantus of Alexandria, *Arithmetics and Book on Polygonal Numbers*, Translated from the ancient Greek by I.N. Veselovsky, ed. and commented by I.G. Bashmakova, Nauka, Moscow, 1974 (in Russian); see also id. edited by Paulus Tannery, *Opera Omnia cum Graecis Commentariis*, Teubner, 2 vols., 1893–1895.

[A4] L. Euler, *Opera Omnia, Ser. I. T. XXI, 91–118*, Leipzig, 1911–1956.

[A5] ———, *Introduction to analysis of the infinitesimals, Book 9, ch. VI.*, Bosquet, Cop., 1988.

[A6] G. Fagnano, *Produzioni mathematiche. Opere Mathematiche del Marchese Giulio Carlo de' Toschi di Fagnano*, see also id., 3 vols., 1911–1913, 1750.

[A7] C. F. Gauss, *Arithmetische Untersuchungen*, Chelsea, New York, 1965.

[A8] ———, *Werke, Bd. 10.*, Göttingen, 1917.

[A9] Ch. Hermite, *Oeuvres,* tome 2, (1) Sur la résolution de l'equation du cinquième degrée, pp. 5–12; (2) Sur l'equation du cinquième degrée, pp. 347–424, Gauthier-Villars, Paris, 1908.

[A10] C. G. J. Jacobi, *Gesammelte Werke*, Band I–VII, Berlin, 1881–1891; Chelsea, New York, ???.

[A11] J. L. Lagrange, Mém. Acad. Berlin (1770–1771).

[A12] A. M. Legendre, *Traite des fonctions elliptiques et des integrales Euleriennes,* t.1–3, Paris, 1825–1832.

[A13] H. Poincaré, *Sur les propriétés des courbes algebriques planes*, J. Liouville **7** (1901), 161–233.

[A14] Serret J.-A., *Cours de calcul différentiel et intégral,* t. 1–3, 2-éme publ., Gauthier-Villars, Paris, 1879.

[A15] E. W. von Tschirnhaus, *Nova Metodus Auferendi Omnes Terminos Intermedios ex Data Aequatione*, Acta eruditorum, Bd. 2, Leipzig, 1683.

B. Main textbooks

[B1] N. I. Akhiezer, *Elements of the theory of elliptic functions*, Amer. Math. Soc., Providence, RI, 1990.

[B2] Z. I. Borevich and I. R. Shafarevich, *Number theory*, Academic Press, New York, 1966.

[B3] N. Bourbaki, *Algèbre,* Ch. 4–17, Masson, Paris, 1981.

[B4] E. Brieskorn and H. Knörrer, *Plane algebraic curves*, Birkhäuser, Basel, 1986.

[B5] W. S. Burnside and A. W. Panton, *The theory of equations*, Dublin Univ. Press, Dublin–London, 1928.

[B6] H. M. Edwards, *Galois theory*, Springer-Verlag, Heidelberg, 1984.

[B7] R. Fricke, *Die elliptischen Funktionen und ihre Anwendungen*, Bd.1, 1916; Bd.2, 1922, Teubner, Leipzig-Berlin.

[B8] P. Griffiths and J. Harris, *Principles of algebraic geometry*, Wiley, 1978.

[B9] A. Hurwitz, *Vorlesungen über allgemeine Funktionentheorie und elliptische Functionen,* Vierte Auflage, Springer-Verlag, Heidelberg, 1964.

[B10] D. Husemoller, *Elliptic curves*, Springer-Verlag, Heidelberg, 1987.

[B11] K. Ireland and M. Rosen, *A classical introduction to modern number theory*, 2nd edition, Springer-Verlag, Heidelberg, 1990.

[B12] M. Koblitz, *Introduction to elliptic curves and modular forms*, Springer-Verlag, Heidelberg, 1984.

[B13] M. Kline, *Mathematical thought from ancient to modern times*, Oxford University Press, New York, 1972.

[B14] S. Lang, *Fundamentals of Diophantine geometry*, Springer-Verlag, Heidelberg, 1983.

[B15] ———, *Elliptic functions*, 2nd edition, Springer-Verlag, Heidelberg, 1987.

[B16] A. I. Markushevich, *The theory of functions of a complex variable*, 3 vols., 1965–1967, Prentice Hall, Englewood Cliffs, NJ.

[B17] L. J. Mordell, *Diophantine equations*, Academic Press, New York, 1969.

[B18] E. J. F. Primrose, *Plane algebraic curves*, Macmillan, London, 1955.

[B19] A. Robert, *Elliptic curves*, Lecture Notes in Math., vol. 326, Springer-Verlag, Heidelberg, 1973.

[B20] J. H. Silverman, *The arithmetic of elliptic curves*, Springer-Verlag, Heidelberg, 1986.

[B21] ———, *Advanced topics in the arithmetic of elliptic curves*, Springer-Verlag, Heidelberg, 1994.

[B22] S. A. Stepanov, *The arithmetic of algebraic curves*, "Nauka", Moscow, 1991. (Russian)

[B23] R. J. Walker, *Algebraic curves*, Dover, New York, 1950.

[B24] H. Weber, *Lehrebuch der Algebra, Bd. 3. Elliptische Funktionen und algebraische Zahlen*, Braunschweig, 1908.

[B25] A. Weil, *Elliptic functions according to Eisenstein and Kronecker*, Springer-Verlag, Heidelberg, 1976.

[B26] E. T. Whittaker and G. N. Watson, *A course of modern analysis*, 4th edition, vol. 1–2, Cambridge Univ. Press, Cambridge, 1927.

C. Selected articles

[C1] B. J. Birch, *Conjectures on elliptic curves*, Theory of Numbers, Proc. Symp. Pure Math., vol. 8, Amer. Math. Soc., Providence, RI, 1963.

[C2] B. J. Birch and H. P. F. Swinnerton-Dyer, *Notes on elliptic curves*, I, II, J. Reine und Angew. Math. **212** (1963), 7–25; **218** (1965), 79–108.

[C3] J. W. S. Cassels, *Diophantine equations with special reference to elliptic curves*, J. London Math. Soc. **41** (1966), 193–291.

[C4] J. Coates and A. Wiles, *On the conjecture of Birch and Swinnerton-Dyer*, Invent. Math. **39** (1977), 223–251.

[C5] F. J. Grunewald and R. Zimmert, *Über einige rationale elliptische Kurven mit freiem Rang* > 8, J. Reine Angew. Math. **296** (1977), 100–107.

[C6] H. Hasse, *Abstrakte Begrundung der komplexen Multiplication und Riemannsche Vermutung in Funktionenkorpern*, Abh. Math. Sem. Hamburg **10** (1934), 325–348.

[C7] L. Holzer, *Minimal solutions of diophantine equations*, Canad. J. Math. **11** (1950), 238–244.

[C8] J. Igusa, *On the transformation theory of elliptic functions*, Amer. J. Math. **81** (1959), 436–452.

[C9] B. Mazur, *Rational points on modular curves*, Lecture Notes in Math., vol. 601, Springer-Verlag, Heidelberg, 1976.

[C10] I. G. Melnikov, *The problem of division of the lemniscate*, I, II, Uchenye Zapiski Leningrad. Gos. Pedagogicheskogo Inst. **197** (1958), 20–38; 39–42. (Russian)

[C11] L. J. Mordell, *On the rational solutions of the indeterminate equations of the third and fourth degrees*, Proc. Camb. Phil. Soc. **21** (1922), 179–192.

[C12] ———, *The infinity of rational solutions of* $y^2 = x^3 + k$, J. London Math. Soc. **41** (1966), 523–525.

[C13] ———, *On the magnitude of the integral solutions of the equation* $ax^2+by^2+cz^2 = 0$, J. Number Theory **1** (1969), 1–3.

[C14] M. Rosen, *Abel's theorem on the lemniscate*, Amer. Math. Monthly **88** (1981), 387–395.

[C15] T. Soundararajan, *On the automorphisms of the complex number field*, Math. Mag. **40** (1967), 213.

[C16] J. Tate, *The arithmetics of elliptic curves*, Invent. Math. **23** (1974), 179–206.
[C17] A. Weil, *Sur une théorème de Mordell*, Bull. Sci. Math. (2) **54** (1930), 182–191.
[C18] P. B. Yale, *Automorphisms of the complex numbers*, Math. Mag. **39** (1966), 135–141.
[C19] D. Zagier, *A one-sentence proof that every prime* $p = 1 \pmod 4$ *is a sum of two squares*, Amer. Math. Monthly **97** (1990), 144.

Index

C_n, 71
$E(\varphi)$, 43
$E(\mathbb{Q})$, the set of rational points on the curve E, 115
$F(\varphi)$, 43
K_n, 72
L-function of a rational elliptic curve, 126
S_p, Serret's curve, 95
Tors $E(\mathbb{Q})$, 124
$\operatorname{cn} u$, 47
$\operatorname{dn} u$, 47
$\equiv$ as identically equal to, 5
η-function, Dedekind's, 160
$\eta(\tau)$, Dedekind's η-function, 160
$\widehat{BM}$, the length of $\smile BM$, 55
$\operatorname{sn} u$, 47
am u, the amplitude, 47
$\wp(z)$, the Weierstrass function, 32
$f(\tau)$, 155
$f_1(\tau)$, 155
$f_2(\tau)$, 155
j-invariant, 101
$j(\tau)$, 169
r_E, 124

Abel, 26, 68, 140
Abel's theorem, 140
algebraic addition theorem, 49
amplitude, 47
automorphisms, trivial, 101
axes, principal, 58

basis of transcendentality, 78
Bernoulli, 68
Bernoulli's lemniscate, 25, 68
Bernoulli, Nicholas, 40
Birch, 125
blow-up, 93
Bring, 147

Cassini, 67
Cassini's ovals, 68
chain, 76
complex multiplication, 101
conic, 1
conjugate (diameters), 58
cubic, 1, 131
cubic, nonsingular, 16
curve, algebraic, 1
curve, elliptic associated with Serret's curve, 97
curve, Serret, 95

Dedekind function, 160
Descartes, 103
diagram, Newton, 99
diameters, conjugate, 58
diophantine equation, 103
Diophantus, 103
division of arcs, 55
division of the arc "by half", 56
domain, fundamental, 173
doubly periodic function, 29
duality, projective, 18
duplication of the arc, 56

Eisenstein, 68, 84
Eisenstein's theorem, 85
element, primitive, 139
ellipse, 25
elliptic integral, 43, 54
elliptic Jacobi functions, 46, 47
equation, cubic, 131
equation, general, 140
equation, modular, 166
equation, quartic, 131
equation, resolving, 136
equation, solvable in radicals, 140
Euclid, 103
Euler, 25, 64, 68
Euler formula, 12
exponential, 50

Fagnano, 25, 60, 68
Fagnano's theorem, 56
Fermat, 104, 118
Fermat primes, 68
field, normal or Galois, 140
form, Legendre, 42
form, Weierstrass, 16, 42
formula, Euler, 12
Fueter, 118
function, doubly periodic, 29

function, elliptic, 29
function, elliptic Jacobi, 31, 47
function, meromorphic, 29
function, Weierstrass, 31, 32
fundamental domain, 173
fundamental parallelogram, 29

Galois field, 140
Gauss, 26, 68
generic lines, 18
generic points, 18
genus of the curve, 98
group of substitutions, 140
group, torsion of the elliptic curve, 124

Hasse, H., 126
Hermite, 131
Hesse curve, 13
Hessian, 13
Holzer, 109

inflection point, 13
integral, elliptic, 43, 54
involution, 89

Jacobi, 26
Jacobi elliptic function, 31

Lagrange's resolvent, 135
Legendre, 26, 60, 107
Legendre form, 42
Legendre's theorem, 41
Leibniz, 40, 145
lemniscate, 67
lemniscate, Bernoulli, 25, 68
lexicographically ordered monomials, 133
lines, generic, 18

Möbius transformation, 159
Main theorem on symmetric polynomials, 133
Mazur, 128
Mazur's theorem, 128
meromorphic function, 29
Mesiriac, 104, 118
Mestre, 125
method of secants, 104
modular equation, 166
Mordell, 118, 119
Mordell's theorem, 119
multiplicity, 93

Newton diagram, 99
normal field, 140
normal form, Weierstrass', 16
normal forms of a nonsingular cubic, 16
number, congruent, 111
numbers, complex, odd, 81

odd complex numbers, 81
order of a point, 114
order of the elliptic function, 31

ovals, Cassini, 68

Pappus's theorem, 5
parallelogram, fundamental, 29
Pascal's theorem, 1, 4
Picard's big theorem, 51
Plücker, J., 7
Poincaré, A., 119
point, fixed, 89
point, inflection, 13
point, nonsingular of the curve, 12
point, simple, 93
point, singular, 93
point, the order of, 114
points, generic, 18
primitive element, 139
primitive root, 137
principal axes, 58
projective duality, 18
projective plane, 9

quartic, 131

rank of the elliptic curve, 124
rationality field, 140
reduction of the curve modulo p, 125
Regiomontanus, 103
resolvents, Lagrange's, 135
resolving equation, 136
resultant, 14
Rosen, 68, 86
Ruffini, Paulo, 140

Serret, 64
Serret's curve, 95
Serret's theorem, 62
Serret, Joseph-Alfred, 61
set, Zorn closed, 76
solution, integer, 103
solution, rational, 103
Steiner, J., 7
sum of points, 1
Swinnerton-Dyer, 125
symmetric, 133
symmetric polynomials, elementary, 133

tangent to the curve, 12
theorem, Abel, 140
theorem, algebraic addition, 49
theorem, Eisenstein, 85
theorem, Fagnano, 56
theorem, Legendre, 41
theorem, Mazur, 128
theorem, Mordell, 119
theorem, on symmetric polynomials, main, 133
theorem, Pappus, 5
theorem, Pascal, 1, 4
theorem, Picard's big, 51
theorem, Serret, 62

theorem, Viéta, 166
theorem, Weierstrass, 50
theta constants, 151
theta functions, 150
trace, 123
transcendentality, basis of, 78
transformation, fractional linear, 159
transformation, Möbius, 159
transformation, Tschirnhaus, 146
Tschirnhaus, E.W. von, 145
Tschirnhaus's transformation, 146

Viéta theorem, 166

Weber, 131
Weierstrass form, 16, 42
Weierstrass function, 31, 32
Weierstrass' theorem, 50
Wiles, A., 126

Zagier, 89

Selected Titles in This Series

(Continued from the front of this publication)

131 **V. V. Sharko,** Functions on manifolds: Algebraic and topological aspects, 1993
130 **V. V. Vershinin,** Cobordisms and spectral sequences, 1993
129 **Mitsuo Morimoto,** An introduction to Sato's hyperfunctions, 1993
128 **V. P. Orevkov,** Complexity of proofs and their transformations in axiomatic theories, 1993
127 **F. L. Zak,** Tangents and secants of algebraic varieties, 1993
126 **M. L. Agranovskiĭ,** Invariant function spaces on homogeneous manifolds of Lie groups and applications, 1993
125 **Masayoshi Nagata,** Theory of commutative fields, 1993
124 **Masahisa Adachi,** Embeddings and immersions, 1993
123 **M. A. Akivis and B. A. Rosenfeld,** Élie Cartan (1869–1951), 1993
122 **Zhang Guan-Hou,** Theory of entire and meromorphic functions: deficient and asymptotic values and singular directions, 1993
121 **I. B. Fesenko and S. V. Vostokov,** Local fields and their extensions: A constructive approach, 1993
120 **Takeyuki Hida and Masuyuki Hitsuda,** Gaussian processes, 1993
119 **M. V. Karasev and V. P. Maslov,** Nonlinear Poisson brackets. Geometry and quantization, 1993
118 **Kenkichi Iwasawa,** Algebraic functions, 1993
117 **Boris Zilber,** Uncountably categorical theories, 1993
116 **G. M. Fel′dman,** Arithmetic of probability distributions, and characterization problems on abelian groups, 1993
115 **Nikolai V. Ivanov,** Subgroups of Teichmüller modular groups, 1992
114 **Seizô Itô,** Diffusion equations, 1992
113 **Michail Zhitomirskiĭ,** Typical singularities of differential 1-forms and Pfaffian equations, 1992
112 **S. A. Lomov,** Introduction to the general theory of singular perturbations, 1992
111 **Simon Gindikin,** Tube domains and the Cauchy problem, 1992
110 **B. V. Shabat,** Introduction to complex analysis Part II. Functions of several variables, 1992
109 **Isao Miyadera,** Nonlinear semigroups, 1992
108 **Takeo Yokonuma,** Tensor spaces and exterior algebra, 1992
107 **B. M. Makarov, M. G. Goluzina, A. A. Lodkin, and A. N. Podkorytov,** Selected problems in real analysis, 1992
106 **G.-C. Wen,** Conformal mappings and boundary value problems, 1992
105 **D. R. Yafaev,** Mathematical scattering theory: General theory, 1992
104 **R. L. Dobrushin, R. Kotecký, and S. Shlosman,** Wulff construction: A global shape from local interaction, 1992
103 **A. K. Tsikh,** Multidimensional residues and their applications, 1992
102 **A. M. Il′in,** Matching of asymptotic expansions of solutions of boundary value problems, 1992
101 **Zhang Zhi-fen, Ding Tong-ren, Huang Wen-zao, and Dong Zhen-xi,** Qualitative theory of differential equations, 1992
100 **V. L. Popov,** Groups, generators, syzygies, and orbits in invariant theory, 1992
99 **Norio Shimakura,** Partial differential operators of elliptic type, 1992
98 **V. A. Vassiliev,** Complements of discriminants of smooth maps: Topology and applications, 1992 (revised edition, 1994)
97 **Itiro Tamura,** Topology of foliations: An introduction, 1992
96 **A. I. Markushevich,** Introduction to the classical theory of Abelian functions, 1992
95 **Guangchang Dong,** Nonlinear partial differential equations of second order, 1991
94 **Yu. S. Il′yashenko,** Finiteness theorems for limit cycles, 1991

(See the AMS catalog for earlier titles)